Environmental Engineering

Series Editors: U. Förstner, R. J. Murphy, W. H. Rulkens

Springer
London
Berlin
Heidelberg
New York
Barcelona
Budapest
Hong Kong
Milan
Paris
Santa Clara
Singapore
Tokyo

D. G. Gorman and J. Neilson (Eds.)

Decommissioning Offshore Structures

With 56 Figures

Springer

Series Editors

Prof. Dr. U. Förstner
Arbeitsbereich Umweltschutztechnik
Technische Universität Hamburg-Harburg
Eißendorfer Straße 40
D-21073 Hamburg, Germany

Prof. Robert J. Murphy
Dept. of Civil Engineering and Mechanics
College of Engineering
University of South Florida
4202 East Fowler Avenue, ENG 118
Tampa, FL 33620-5350, USA

Prof. Dr. ir. W.H. Rulkens
Wageningen Agricultural University
Dept. of Environmental Technology
Bomenweg 2, P.O. Box 8129
NL-6700 EV Wageningen, The Netherlands

Editors

Prof. D. G. Gorman
The Oil & Gas Institute
The University of Aberdeen
Aberdeen AB24 3UE, UK

Dr. J. Neilson
The Oil & Gas Institute
The University of Aberdeen
Aberdeen AB24 3UE, UK

ISBN-13: 978-1-4471-1554-0

British Library Cataloguing in Publication Data
Decommissioning offshore structures. - (Environmental engineering)
1 Offshore structures
I.Gorman, D. G. (Daniel Geelan), 1949- II.Neilson, June
627.9'8 ISBN-13: 978-1-4471-1554-0

Library of Congress Cataloging-in-Publication Data
Decommissioning offshore structures / D.G. Gorman and J. Neilson (eds.)
p. cm. -- (Environmental engineering)
ISBN-13: 987-1-4471-1554-0 e-ISBN-13: 987-1-4471-1552-6
DOI: 10.1007/987-1-4471-1552-6
1. Drilling platforms. 2. Offshore structures. 3. Oil wells--Abandonment.
I. Gorman, D.G. II. Neilson, J. (June), 1948- . III. Series: Environmental engineering (Berlin, Germany)
TN871.4.D43 1997 97-29240
622'.33819--dc 21 CIP

Softcover repint of the hardcover 1st edition 1998

Typesetting: Camera ready by the editors

69/3830-543210 Printed on acid-free paper

Preface

This book is an extremely important benchmark in the development of a co-ordinated approach and a framework for a methodology of decommissioning offshore installations on the North West European Continental Shelf, and I am pleased and grateful that Professor Gorman and Dr Neilson have asked me to contribute this preface to the volume.

My contribution, requested by the University of Aberdeen, was to have been only the chapter entitled "Technical Aspects of Decommissioning Offshore Structures". Little did I know that the consequences for me in agreeing to prepare a paper for a conference on the subject of Decommissioning in September 1996 would be the beginning of a dialogue and interaction which would ultimately lead to my appointment in July 1997 as Co-ordinator of the Oil & Gas Institute of the University of Aberdeen. One of the major reasons for this development has been my interest in, and increasing awareness of, the complexity and scale of the task that is facing the UK and Norway (in particular) over the next two decades in decommissioning the infrastructure of platforms, pipelines and sub-sea equipment in a manner which meets the demands and aspirations of legal, political, economic, environmental and engineering interest groups.

This book is the first attempt to bring together a coherent body of specialist knowledge in those principal areas and has resulted in a clear exposition of the high degree of interdependence of knowledge in those areas necessary for the development of the so-called Best Practicable Environmental Option in each case.

It is the case that the University of Aberdeen, with its general strength and particular experience over the last twenty-five years, is usually well placed to pull together the multi-disciplinary strands of Law, Economics, Engineering, Geosciences and Environmental Management to develop a national strategy and core of expertise for decommissioning, disposal and removal of the multi-million tonne residue of the offshore oil and gas industry as it moves inexorably from its present mature phase to the inevitable post-mature and depletion phase.

All eyes have been fixed on the Shell Brent Spar story as it has progressed, and all minds are pre-occupied with the lessons that have been learned very swiftly in the last two years. These lessons have changed very rapidly the conventional wisdom of potential removal and disposal options and methods with the result that the Oil and Gas Institute of the University of Aberdeen wish to further develop knowledge in this area, and to disseminate as rapidly as possible that knowledge.

The publication of this book by Springer-Verlag is very timely and must be welcomed by all who are, or will soon be, faced with the responsibility of managing and engineering major decommissioning projects. I sincerely hope that the pace of research and study in the area of decommissioning within universities, oil company operators, design contractors and by the supply companies, from whom the industry can expect continued innovation, will result in further published material.

Finally, I take this opportunity to congratulate the editors, Professor Gorman and Dr Neilson, for the professional manner in which they co-ordinated the contributions for such a wide and distinguished group of academics, and to Lisa Robertson for her patience and thoroughness during the final preparation of the manuscript.

Peter A Meenan
Coordinator, Oil & Gas Institute
The University of Aberdeen
June 1997

Contributors

G Simpson holds the Schlumberger Chair in Energy Industry Management at the University of Aberdeen. He worked for Esso/Exxon for over 20 years in the UK, the USA and Norway, in various technical and managerial roles, and joined the University of Aberdeen in early 1997. He has a PhD in Geology and an MBA and is a Fellow of the Geological Society, a member of The Institute of Petroleum, The Society of Petroleum Engineers and The Petroleum Exploration Society of Great Britain.

P A Meenan is co-ordinator of the Oil and Gas Institute of the University of Aberdeen. His chief areas of experience lie in the Project Management, Design, and Construction Supervision of civil/structural and marine engineering contracts including steel bridges and offshore structures. Experience has included quality assurance of welding; large diameter piling on and offshore and heavy reinforced concrete foundations; design and construction of major temporary works; offshore structure loadout operations and offshore construction/installation, including very heavy lifting operations; management of advanced F.E. analysis projects.

A G Kemp is currently Professor of Economics at the University of Aberdeen. For many years he has specialised in his research in petroleum economics with special reference to licensing and taxation issues. For many years he has been a consultant on petroleum contracts and legislation to a large number of governments, the World Bank, the United Nations, oil companies and the Commonwealth Secretariat. He was a specialist adviser to the UK House of Commons select Committee on Energy from 1980 until its abolition in 1992. He is member of the Secretary of State for Scotland's independent panel of economic consultants. He is an editorial adviser to a number of energy and other academic/professional journals. In 1993 he was appointed by the Minister of Energy to the new UK Government Independent Energy Advisory Panel. He is Director of Aberdeen University Petroleum and Economic Consultants (AUPEC) which provides consultancy service in petroleum services.

L Stephen is a Research Fellow in Petroleum Economics at Aberdeen University. Since 1994 she has worked on a number of papers and articles on prospects, decommissioning and related issues in the UKCS with Professor A G Kemp.

M J Baker is the first holder of the U.K.'s first Chair of Safety Engineering which was established in 1990 at the University of Aberdeen. Prior to that he was a lecturer and then Reader in Structural Reliability at Imperial College, London. He has published widely in the area of structural integrity and serves on a number of national and international committees concerned with safety and reliability. He is a chartered civil and structural engineer and a Fellow of the Safety and Reliability Society.

A D M Forte was appointed to the Chair of Commercial Law in the University of Aberdeen in 1993, having taught previously at the universities of Glasgow, Dundee and Edinburgh. His published work has concentrated mainly on commercial contracts, insurance law and, more recently, on salvage. Professor Forte is a member of the Banking Law Committee of the Law Society of Scotland and of the Advisory Committee on Contract Law to the Scottish Law Commission.

G Jordan is a Professor of Politics in the Department of Politics and International Relations, University of Aberdeen. His main research interests are in British central government and in interest group theory. He has also worked on budgetary issues and local government. In 1992 he published a study of the Finniston Committee entitled *Engineers and Professional Self Regulation*. His latest book (with William Maloney) is *The Protest Business? Mobilizing Campaign Groups*, Manchester University Press 1997.

L Bennie is a Lecturer in Politics in the Department of Politics and International Relations, University of Aberdeen. Her research interests are in green politics and Britain's political parties. Recent publications include (with J. Mitchell and J. Brand) *How Scotland Votes: Scottish Parties and Elections*, Manchester University Press 1997. She is currently completing a PhD thesis on the Scottish Green Party.

D Gorman is a Professor in Engineering at the University of Aberdeen. His major field or research is structural analysis, and in particular vibration analysis. He is a Fellow of the Institution of Mechanical Engineers and the Institution of Engineers and Shipbuilders in Scotland. Currently, he is course director for the Aberdeen University MSc programme in Technology and Management in the Oil and Gas Industry.

J Neilson is employed in the Department of Politics and International Relations, University of Aberdeen. Prior to completing her PhD in 1990, she was a Librarian at Queen Mary and Westfield College, London where she specialised in British, European and International Standards relating to social and environmental issues.

Contents

1

The Environmental Management Aspects of Decommissioning Offshore Structures

G Simpson

Introduction

This chapter reviews the environmental management aspects of the decommissioning of offshore structures, with particular reference to the United Kingdom Continental Shelf. After some brief background material and a discussion of the principles involved, the topic is described in terms of three separate levels. Level one is defined by the minimum requirements, as set out in the legal framework and government guidelines and requirements. The second level is represented by a set of guidelines, agreed by the operating companies and published by the association that represents them, the United Kingdom Offshore Operators Association (UKOOA). These go somewhat beyond the legal minimum, and provide a management system and an environmental checklist for all offshore operations, of which decommissioning is just one. Finally, an examination of the specific case of Brent Spar will illustrate how Shell fulfilled all requirements to the letter, went some way beyond even UKOOA's guidelines, and still encountered major difficulties. The management issue at this point goes beyond the quantitative ticking off of tasks and conditions, and into the softer qualitative areas of politics and sociology, social responsibility and ethics.

Background

Approximately 220 oil and gas facilities and about 5,000 kilometres of pipeline have been built and installed on the United Kingdom Continental Shelf (UKCS) since the mid 1960s, with similar numbers applying to the combined totals of the other sectors of the North Sea, in Norway, Denmark, the Netherlands and Ireland [1]. The installations cover the full range of types, including platforms made of either steel or concrete, which sit on the sea-bed, floating production systems, often including offshore storage and loading facilities (FPSOs), other offshore storage and loading units (e.g. Brent Spar and Fulmar FSU), sub-sea production systems and pipelines. Each is different, and presents its own problems in terms of water depth, type and size. In the shallower waters of the southern North Sea, platforms typically weigh less than 5,000 tonnes; some deep water steel platforms are an order of magnitude heavier than this and the largest concrete platforms weigh over 200,000 tonnes. The Norwegian Troll platform is the largest moveable man-made structure ever built, at over one million tonnes displacement [2].

Since even the smallest and most mobile installations handle substances which could have an impact on the environment, and they are all situated in a sea which has many other social and commercial uses and users, the environmental sensitivity to decommissioning arrangements is considerable. However, this needs to be put in context. Decommissioning is an integral and, one could argue, on a net present value basis, a relatively small part of the oil and gas production cycle. But as older fields progress from plateau into decline, new production is needed to maintain supplies. This requires investment, and, in principle, at least, the less is spent on decommissioning, the more is available for exploration and appraisal. Furthermore, since expenditure on decommissioning can be offset against tax, decommissioning bills are in part paid by the taxpayer.

Principles

In determining the "best" strategy for decommissioning an installation, a balance has to be struck between technical difficulty and safety, the impact on the environment and costs.

In terms of impact on the environment, this needs to be considered in light of:

- the inventory of materials forming and contained within the installation,
- how quickly, if at all, these materials will be released into the environment, land, sea or air, by the various decommissioning strategies being considered,
- the energy requirements or savings of each possible solution, and the effects of the related atmospheric emissions,
- the characteristics of the environments which could be affected and their inhabitants,
- the extent to which each of the environments will be affected

For example, the inventory of materials will need to consider:

- the structure, concrete and/or steel, and perhaps haematite ballast, all of which are biologically inert, but also small quantities of materials such as paint, perhaps containing zinc, sacrificial anodes, containing aluminium and zinc, and perhaps also bismuth, iron, indium, titanium, cadmium, lead and mercury, biocides in ballast water, and hydrocarbon residues.
- topsides, containing a wide variety of materials including wood, plastics, glass, electrical wiring, motors, batteries, cables, asbestos, oil and lubricants and pipes containing residues from production, including Low Specific Activity (LSA) scale.

- piles of drill cuttings, small fragments of rock produced by the drilling process, mostly coated in drilling mud, some of which is oil based, some water based. The ALTRA report for UKOOA [9] assesses the environmental implications of these piles, which have been observed around approximately 60 platforms in the Northern and Central North Seas. It seems that the wave and tidal action is sufficient to disperse the cuttings around the shallow water Southern North Sea platforms. Some of these piles, which can have volumes as large as 25,000 cubic metres and be up to 20 metres high, have been lying on the sea-floor for many years.

We need to consider, at one extreme, the effects (some negative, some perhaps positive, see Picken in reference 3 pp. 213-222) of dumping or leaving all these materials in the marine environment; at the other, the technical and safety problems, and the effect on land and air, of bringing an installation ashore for dismantling, offset by the potential benefits to the environment of re-cycling and hence energy savings in not having to produce from new raw materials.

In order to assist in the assessment of each of these issues, and of the appropriate balance and hence the optimum solution, various sets of laws, guidelines and practices have been formulated, and these are summarised below.

The decommissioning of pipelines is not a major issue at the moment. The small diameter infield lines can easily be removed, when that or those fields cease production, by a reversal of the installation method. The larger diameter trunk lines, which would be much more difficult to retrieve, are likely to have long and useful lives as they facilitate the development of smaller fields which, without being able to transport their products through these trunk lines, might not be economic. The same could well be true, of course, for many platforms, deferring their decommissioning considerably.

Level 1: The Legal Framework

The details of the Legal Framework governing decommissioning will be discussed in detail in a subsequent chapter, but they are summarised here for the sake of completeness. At the start of the UKCS exploration activity, it was assumed by many [3, p26] that all installations would be removed entirely and that the marine environment would be restored to its original condition, as prescribed in the 1958 Geneva Convention [4]. However, subsequent and overriding legislation, including the United Nations Law of the Sea Convention and the IMO Guidelines (see below), introduced the possibility of partial removal.

The Petroleum Act 1987 sets out the requirements for the decommissioning of offshore facilities on the United Kingdom Continental Shelf (UKCS). The Department of Trade and Industry (DTI) is charged with ensuring that the requirements of the Petroleum Act are complied with, and therefore companies are required to submit to the Department's Abandonment Unit in the Oil and Gas Division in Aberdeen, for approval, an Abandonment Programme.

Government policy, as set out in the DTI's guidelines [7], from which the following statements are extracted (with my italics emphasising those parts which relate to environmental impact), is that:

- Government will seek, via a case-by-case approach, to achieve effective and balanced abandonment solutions which are consistent with international obligations, and have a proper regard for the environment, other uses of the sea and cost considerations.
- The Department will seek to ensure:
    - that interested parties have a clear view of the policy and procedures;
    - that decisions on abandonment are based on full information, are taken in a speedy and efficient manner and place as little administrative burden as possible on the various parties concerned;
    - that appropriate mechanisms exist for the sharing and transfer of information and provide for proper account to be taken of experience gained and lessons learned.

In considering each Abandonment Programme, the DTI also takes into account the requirements of other UK legislation and of international law. Thus, before the decommissioning of an offshore installation can commence, the following must be obtained or provided, as appropriate:

- approval of an abandonment programme under the Petroleum Act 1987;
- *approval of a well abandonment programme* in accordance with the obligations contained in the licence;
- acceptance of a Safety Case under the Offshore Installations (Safety Case) Regulations 1992;
- *confirmation that the requirements of the Coast Protection Act 1949 have been satisfied.*

If any operation or final disposal on-shore is proposed:

- compliance with the relevant health, safety, pollution-prevention and waste-regulation requirements including, in particular, Part II of the Environmental Protection Act 1990.

If marine disposal is proposed:

    - a licence under the Food and Environment Protection Act 1985;
    - authorisation under the Radioactive Substances Act 1993 in certain circumstances.

To expand specifically on each of these agreements and Acts:

Current Licences contain a clause which requires a licensee to obtain written consent from the Secretary of State before a well is abandoned. Well consents are currently administered by the Health a Safety Executive on behalf of the DTI.

The Coast Protection Act 1949 states that the consent of the Secretary of State is required before an installation can be placed on the UKCS, and this consent usually contains a condition relating to the decommissioning of these facilities, to the effect that no obstruction or danger to navigation should be caused by or be likely to result from the decommissioning activities.

The Environmental Protection Act 1990, which governs the onshore disposal of material, introduced a system of integrated pollution control for specified processes to prevent, minimise or render harmless any releases to the environment and imposed a "duty of care" on those involved in the disposal of any wastes.

Under the **Food and Environment Protection Act 1985**, a licence is required to deposit substances offshore, either in the sea or beneath the seabed. The criteria for the issuing of a licence involve protection of the marine environment, the living resources which it supports and human health, and the prevention of interference with legitimate uses of the sea.

The Radioactive Substances Act 1993 controls the keeping and use of radioactive materials and the disposal and accumulation of radioactive wastes. It applies to offshore installations and to a specified area around such installations.

From an international perspective, the International Maritime Organisation (IMO), an agency of the United Nations, is widely recognised as the "competent international organisation" [3 p8], and therefore the DTI's guidelines have been drafted such that they are consistent with the IMO's 1989 Guidelines and Standards [6]. For example, there is the requirement for the complete removal of all installations weighing less than 4000 tonnes in air and in water depths of less than 75 metres. For installations emplaced on or after 1 January 1998, the minimum depth rises to 100 metres. After 1 January 1998 only those structures whose designs make complete removal technically feasible may be installed. For installations which are in greater than 75 metres of water, or which weigh more than 4000 tonnes in air, partial removal is permitted, provided a minimum of 55 metres of clear water remains above any submerged remains, and that the remains are maintained to prevent structural failure. If complete or partial retention of an structure would enhance marine life, or serve some other use, then permission may be given for it to be left in place.

Approximately 75% of UKCS structures are in shallow water, are relatively small, and must be removed entirely, although deep sea disposal remains an option. This removal will be relatively straightforward, with essentially no impact on the environment and, indeed, this has proved to be the case for the nine such structures already decommissioned (see later). For the other 25%, the largest platforms, the decommissioning options include complete removal, partial removal (including toppling), deep sea disposal, and leaving the installation in place. The regulations require that in formulating and presenting for approval a preferred decommissioning option for each of these installations, Operators should carry out an assessment which leads to the identification of the "Best Practicable Environmental Option" (BPEO). The BPEO is the option which minimises the impact of all decommissioning activities on all environments at an acceptable cost in financial and safety terms, recognising, as stated above, that it is in the interests of both the nation and the industry to minimise decommissioning costs as far as possible, without compromising safety and environmental considerations, and that a balance acceptable to "the broad community" has to be reached. Until recently it had been assumed that the Government, through the DTI, had a mandate to be the arbiter of what constitutes an acceptable balance. If there is going to be a management problem, it lies in the potential for parties with different interests and perspectives to have different views as to the "correct" balance, and hence the "right" decommissioning solution.

The IMO Guidelines [6] also propose that the assessment of any potential to affect the marine environment should be based upon "scientific evidence, taking into account the effect on seawater quality, geological or hydrographic characteristics, the presence of endangered or threatened species, existing habitat types, local fishery resources and the potential for pollution or contamination of the site by residual products from, or deterioration of, the offshore installation or structure." Decommissioning should be planned and carried out in such a way as to cause no significant adverse effects upon navigation or the marine environment, and its inhabitants, especially threatened or endangered species.

The UK Government is a contracting party to several conventions which relate to the prevention of marine pollution. These include the Oslo Convention 1972, the London Convention 1972, the Paris Convention 1974, the United Nations Convention on the Law of the Sea 1982, and the OSPAR Convention 1992. The London and Oslo Conventions control the disposal of substances at sea, London applying to all marine waters world-wide, Oslo applying only to specified areas in the North East Atlantic, including the North Sea and parts of the Arctic Ocean. Both prohibit the dumping of substances or materials without the approval of the relevant authority, and the Oslo Convention requires that bulky wastes, which may represent a hazard to fishing or to navigation, be dumped in greater than 2000 metres of water and greater than 150 nautical miles from land. The OSPAR Convention includes specific reference to the dumping of disused offshore installations and pipelines, and when ratified, will replace the Oslo Convention.

The DTI Guidelines [7] also set out a four stage process (preliminary discussions, detailed discussions, implementation and monitoring) for handling decommissioning programmes, and they set out in some detail a recommended format for individual proposals. An appendix to the guidelines contains a list of those organisations to be consulted during the preparation of proposals.

Level 2: UKOOA Guidelines

Those companies operating installations on the UKCS, all being members of UKOOA, have agreed a set of guidelines for governing all of their offshore operations in line with generally accepted best environmental practice. Decommissioning is just a specific instance in which these general guidelines should apply. The guidelines were originally agreed by the UKOOA Council at its meeting on August 14 1991, and have been published by UKOOA [9]. It is worth setting out these guidelines in some detail, as it illustrates the depth of the commitment of UKCS operators to the environment, and it describes the management methodology used to fulfil this commitment. The guidelines are therefore summarised below:

The purpose of the guidelines is to form a framework within which operators can conduct and audit their business in a socially responsible and ethical manner that protects safety, health and the environment. The guidelines reflect the significant role which contractors play in the industry in that they specifically address the effects of operations, often carried out by contractors, on the environment and the natural processes within that environment.

Over and above continuing compliance with all applicable UK environmental legislation and good oilfield practice, each member company of UKOOA is committed to managing its business according to the following principles. These principles will be given a high priority in the definition and implementation of corporate strategies, and in developing environmental management systems appropriate to each member's overall operation:

- adopt appropriate procedures, industry practices and other operating guidelines with the objective of protecting the environment.
- promote among employees and contractors an individual and a collective sense of responsibility for protecting the environment, incorporating the principle of line management responsibility.
- endeavour to conduct operations and handle all materials, products and waste in a manner that protects the environment and conserves energy. Provide advice to suppliers, contractors and others where necessary.
- develop and maintain procedures to minimise the risk of all accidental spills and emissions and maintain appropriate emergency response procedures in case of accidents.
- develop, implement and monitor programmes to manage overall emissions and waste generation and disposal to minimise the environmental impact.

- co-operate with Government and European Authorities to help develop laws and regulations and review issues to safeguard the environment.
- support research and development programmes to study the effects of the industry's activities on the environment and any associated risks, leading to improvements in environmental protection practices.
- contribute to the environmental debate and discussion pertaining to the offshore industry.
- undertake appropriate internal reviews and evaluations of operations to measure progress and ensure compliance with the above principles.

UKOOA members are committed to conducting their operations with high priority being given to environmental protection. The specific management processes are encapsulated in an Environmental Management System (EMS) which:

- should be based on the philosophy and principles of quality management.
- should consider both the design and operation of facilities, the contribution of both human and hardware factors in environmental management and incorporate the Best Available Technique Not Entailing Excessive Costs (BATNEEC) principle.
- should be comprehensive in its scope, covering the environmental protection of the atmosphere, water and the land.
- would be most effective in a proactive goal-setting regulatory framework (rather than a highly prescriptive one), which allows and encourages efforts to develop new ways to improve performance.

It must be emphasised that in the development of an EMS there are no set formulae. Its contents will depend upon the size and culture of individual companies, the nature and maturity of their operations, management organisation and philosophy. An EMS is defined as the organisational structure, objectives, procedures, processes and resources for implementing effective management with an integrated systematic approach to environmental protection.

The ultimate responsibility for, and commitment to, a company's environmental policies and objectives rests with senior management. However the aim must be to have the EMS "owned" by all personnel, so that they share in the responsibilities, understand the limits of their authorities, and are fully aware of their accountability and the criteria by which they will be judged. The development of the EMS should therefore involve the personnel who will have to apply it, thus achieving "ownership". Management must actively, visibly and frequently spend time motivating, encouraging, and increasing involvement and awareness. Management must also ensure that the systems developed and implemented by any contractor an Operator uses should be compatible with the Operator's EMS.

Each Operator must develop its own well documented system of policies and standards which are demonstrably followed. The process of internal audit and review can then be used to show that the EMS is being implemented. This will enable the Operator to improve internal commitment to the system and to demonstrate to the authorities that it practices sound environmental management. This in turn ought to minimise the need for prescriptive legislation.

An EMS which is based on the philosophy and principles of quality management should include a number of key generic processes, including:

- a policy and standards process to promulgate the company's environmental aims, principles and objectives to be achieved by all staff including contractors.
- an activity analysis process to ensure that controls appropriate to the nature and significance of an activity are systematically applied.
- a formal process to assign authority for clearly defined sets of activities to the appropriate competent persons in the organisation.
- a monitoring and audit process to provide assurance that the EMS is functioning correctly, and to train and assess the performance of those responsible for achieving specific objectives.
- a follow-up process to ensure that rectification actually occurs.

The EMS documentation must include the following items:

- a description of the EMS Structure
- a statement of policy and objectives
- the organisational structure and responsibilities
- the qualifications, recruitment and training requirements for operator and contract staff at all levels, both offshore and onshore, such that the focal points for environmental responsibility are defined
- a statement on employee involvement and motivation
- an environmental assessment of the facilities including baseline studies and waste minimisation
- contingency planning and emergency response plans
- procedures for reporting, investigation and follow up of environmental incidents
- procedures for performance monitoring, measurement and reporting, including quality control, compliance with company standards and legislation
- procedures for internal audit, review and follow up

- procedures for systematic re-appraisal and follow up
- performance of contract operations
- **decommissioning and restoration**

The most important objectives in implementing an EMS must be to provide a means of controlling the impact on the environment of its operations, and a system which encourages continuous improvements in performance. The EMS establishes the way in which a company's environmental performance will be judged by employees, customers, the authorities and the public. It is in all companies' interests therefore to ensure that the EMS is a realistic, continually evolving process and that it is understood, accepted and followed by all concerned.

The UKOOA Guidelines [9] then go into considerable detail in specifying what its members mean in terms of environmental training, environmental assessment, performance monitoring, audits etc.

The significance of these UKOOA guidelines is that each UKOOA member in effect commits, through its membership, to adopt these guidelines and to conduct its operations in a manner which is demonstrably, via audit, consistent with them. The guidelines themselves are consistent with the Legal Framework, but in many areas go beyond them, while also providing a management system for their practical implementation.

Perhaps the most comprehensive attempt to evaluate the relative impact on the environment of the various decommissioning options is that set out in a report by AURIS for UKOOA [8]. A summary of this was presented by Picken to the House of Lords Select Committee [3, pp 213-218]. The essence of this report seems to represent the "industry view", and can be summarised as follows.

- the environmental impacts of the options allowed under IMO guidelines are all very small, and therefore it is difficult to discriminate between the various options on environmental grounds alone. Thus considerations of "technical feasibility, human safety risk and financial cost" are likely to determine the "preferred option" or BPEO.
- the study, nevertheless, does attempt to identify the (marginally) preferred options for topsides, steel jackets, concrete gravity bases, cuttings piles and pipelines, making judgements on the relative impacts, in the short and longer terms, on "water quality" versus "benthic ecology" and so on.

Level 3: Industry Experience and Practice

The DTI's "Brown Book" lists nineteen UKCS installations which have been decommissioned up to the end of 1996 and for which an approved abandonment programme exists under the Petroleum Act, 1987 [10 and Table 1]. Four of these were floating production systems (Crawford, Argyll, Angus and Emerald) and two were subsea systems (Blair and Staffa). Each of these required only the capping of wells and tidying up of the seabed, which was achieved with no predicted or actual environmental impact and at relatively low cost (an average of about £5M per installation) [14]. Nine installations were fixed steel platforms, in shallow water, at Forbes, Esmond, Gordon, Viking and Leman. These were completely removed, at a somewhat higher cost (an average of about £10M each) [14], but again with no predicted or actual environmental impact. Four of the remaining five installations were redundant parts of larger installations (Brent Spar, Fulmar FSU and SALM, Frigg Flare Column) and the last is Piper Alpha, where there were special constraints, as discussed below. Thus, up to the end of 1996, the UK has little practical experience of large structures, nor is there much experience in other parts of the world that relates directly to the size of structure and the environmental conditions of the UKCS.

The list of UKCS installations decommissioned as of March 1997, and hence the sum total of our experience, is as in Table 1.

UKCS Installations Decommissioned

Data from the DTI's "Brown Book" [10]

Piper A: a fixed steel platform in 145 metres of water, abandoned in 1988 after being destroyed by an explosion and fire.

Crawford: a floating production system, removed in 1991

Argyll, Duncan, Innes: a floating production system, removed in 1992

Blair: a subsea production system, removed in 1992

Forbes: a fixed steel platform, in 34 metres of water, totally removed in 1993

Angus: a floating production system, removed in 1993

Fulmar SALM and FSU: a mooring buoy and a storage unit, each removed in 1994

Brent Spar: an oil storage and loading facility removed in 1994, but not yet decommissioned

Esmond: two fixed steel platforms, in 34 metres of water, totally removed in 1995

Gordon: a fixed steel platform, in 34 metres of water, totally removed in 1995

Viking: four fixed steel platforms, in 26 metres of water, totally removed in 1996

Leman BK: a fixed steel platform, in 37 metres of water, removal in progress

Frigg FP: a flare column totally removed in 1996

Emerald: a floating production system, removed in 1996

Staffa: a subsea production system, removal in progress

Table 1. UKCS Installations Decommissioned (as of March 1997)

Two of the installations listed in Table 1 are of particular interest.

For Piper Alpha, a detailed assessment of all the possible courses of action resulted in a decision to topple the remains in situ, on the grounds that they were so badly damaged that to attempt any other solution would be too dangerous for the workforce [3, p164]. No significant environmental impacts resulting from this have been observed. Indeed, it seems to be clear, from our experience with sunk shipping, not only that millions of tonnes of steel has no real impact on the marine environment, but that even if that steel is heavily contaminated with hydrocarbons, the effect is so diluted as to have negligible impact.

Brent Spar is by far the best documented of recent decommissioning efforts, (eg, reference 11, from which most of the following is taken) and provides many lessons for the future. The chronology and background to the environmental discussion, and the rationale for the decisions taken, is therefore given in some detail below, not to push any particular view or to deliver a verdict on the various arguments put forward, but to illustrate the process of decommissioning and its complexity, in a real situation.

The Brent Spar was commissioned in 1976. It is a 29 metre diameter cylindrical buoy which floats vertically in the water with a draft of 109 metres and a height above water of 28 metres. It was constructed using techniques similar to those of ship building and consists of a thin outer skin of 20 mm plate steel, stiffened by ribs and bulkheads.

The total weight of the structure is 14,500 tonnes, made up of 6,700 tonnes of structural steel, 6,800 tonnes of permanent haematite and concrete ballast and 1000 tonnes of equipment. The upper section consists of a helideck, crane, tanker mooring and loading boom and accommodation for 30 people.

The top of the lower section is made up of twelve buoyancy tanks. Below this, the main storage area is divided into six tanks which extend to the base of the buoy. The total storage capacity is 300,000 barrels of oil. At the base of the buoy there is a sealed compartment containing the permanent ballast, which is composed of haematite embedded in concrete.

During operations, the buoy was used to store oil from Brent Alpha platform, and it also acted as a tanker loading facility for the whole Brent Field. It was the sole route for the export of crude oil until the Brent System Pipeline was commissioned in 1978. After this, it continued to be used as an alternative to the pipeline system.

The Brent Spar was held on location by a six leg catenary mooring system. Each mooring line was made up of a 285 metre length of 4 inch chain, one end of which was attached to the buoy and the other to 800 metres of 3.5 inch wire, connected directly to an anchor block.

A manifold, which acts as a junction in the pipeline between the Brent Alpha and Beta platforms, is located on the sea bed directly below the Spar. This is now isolated and will remain in place, operating as a fully maintained sub-sea facility.

The water depth at the Spar location is 140 metres.

During a nine day operation in 1976 between construction and installation, the Spar was upended by the gradual ballasting of the storage tanks. Despite the care taken during this operation, the buoy was overstressed to some extent by the pressures it experienced. In January 1977, two of the main storage tanks were ruptured by the accidental build up of differential pressures, which were in excess of design limits, between the inside and outside of the tanks. The subsequent repairs were conducted only to maintain structural integrity of the buoy not to make the tanks water tight. The tanks were not used again for oil storage and remained filled with sea water.

In 1978, the Brent System Pipeline to Sullom Voe Terminal was commissioned and this became the main export route for the Brent crude. After this, the Spar was used only as an alternative export route.

Between 1987 and 1990, the maintenance costs of the facility increased substantially, and in 1991 it was concluded that further investment of over £90 million was necessary if the buoy's life was to be extended and that in this case the facility would be out of commission for two to three years. This was deemed to be an economically unattractive investment and it was concluded that the structure should be decommissioned.

The Brent Spar was taken out of commission in October 1991, at which point the undamaged storage tanks were emptied of crude oil and filled with sea water. The process pipework was flushed through with sea water and the storage tank contents were pumped into a shuttle tanker. All the buoyancy tanks were emptied and all the valves, watertight hatches and doors were shut to prevent flooding. All items of loose equipment, including fire fighting equipment, life saving appliances and spares were removed. In November 1991, the manifold was isolated and the flexible risers were removed.

The Brent Spar is different from any other installation on the UKCS, and therefore, whatever method of decommissioning was chosen for it, was not considered to have set a precedent for the disposal of other structures, most of which, as stated earlier, will be brought ashore and dismantled.

Decommissioning studies began in October 1991, and in September 1992 Shell first discussed the various options with the regulatory authorities. Thirteen possible methods of abandoning or re-using the buoy were initially put forward for consideration, of which the following six were identified as viable options:

- Horizontal Dismantling (and Onshore Disposal)

- Vertical Dismantling (and Onshore Disposal)
- In-Field Disposal
- Deep Water Disposal
- Refurbishment and Re-Use
- Continued Maintenance

Of these, horizontal dismantling and deep water disposal were considered in greater detail.

Horizontal dismantling presented a fundamental engineering challenge. Raising the Spar in the water without posing undue risk to people or the environment would be very difficult. While the Spar was still fit for the purpose for which it was designed, estimates of its structural strength under various stresses indicated that the original installation process could not simply be reversed without a significant risk that the Spar would break its back. Its design required that a balance of internal and external pressures be maintained in order to stop the tank walls from rupturing. The risks here are increased by the fact that two of the Spar's six storage tanks have been damaged.

Decommissioning studies continued through 1993, including thirty independent studies, and in February 1994 a study commissioned through Aberdeen University endorsed deep water disposal as the preferred option. This was followed in late 1994 by BPEO and Impact Assessment studies by Rudall Blanchard Associates [12, 13] which also concluded that deep sea disposal was the most sensible alternative, and that there would be no significant environmental impact.

During the BPEO identification process, Shell reported regularly to the Department of Trade and Industry (DTI). When deep sea disposal began to emerge as the preferred option, the DTI requested a thorough review of the issues, to be carried out by staff at the Marine Laboratory Aberdeen, part of the UK Government's Scottish Office. One of the outputs of this was that in the case of onshore scrapping, the probability of a fatality was six times greater and the cost four times greater with no environmental benefit [11]. The environmental risks of onshore decommissioning were considered to be marginally greater than those relating to deep-sea disposal. The clear balance in favour of deep sea disposal resulted in the decision not to carry out further detailed analysis of the onshore option. It is perhaps worth noting here that UKOOA estimates [3, p.111] the cost of recovering offshore steel to be between £3000 and £4000 per tonne, and its resale value as scrap to be between £50 and £100 per tonne, so there is no obvious economic incentive for onshore scrapping.

Specifically, the studies showed that the relatively small amount of contaminants on and in the Brent Spar would have only a small and localised impact on any environment into which they were discharged or released.

Deep sea disposal would have a small localised impact at the deep water disposal site but no effect on the coastal or onshore environment. If all were to go as planned, horizontal dismantling of Brent Spar would have had negligible impacts on the marine environment, and any effects onshore would be extremely localised. However, there is potential for an unplanned event during horizontal dismantling, and if this were to occur in shallow inshore waters there could be a significant impact on other users of the sea. The environmental impacts of each option were therefore judged to be evenly balanced.

As it became clear, during late 1993 and early 1994 that deep sea disposal was the preferred option, Shell formally consulted with, and sent out information and documentation to, a wide range of interested parties including members of the Oslo and Paris Commission (OSPAR), the media, and Greenpeace. No adverse response was received and there was little public or media comment.

The following bodies were consulted:

> Scottish National Heritage, Scottish Fishermen's Federation (SFF), Joint Nature Conservancy Committee (Seabirds at Sea), Orkney Fishermen's Association, Firth of Forth Fishermen's Association, Federation of Highlands and Islands Fishermen, Western Isles Fishermen's Association, British Telecom International, the Scottish Office Agriculture and Fisheries Department, Her Majesty's Industrial Pollution Inspectorate and the Health and Safety Executive. Other interested government bodies were also notified of the Abandonment Plan, including the Ministry of Defence, Hydrographer of the Navy, Crown Estates Commissioners for Scotland and the Department of Transport.

Again, no objections were raised to Shell's recommendation that deep sea disposal was the preferred option for abandoning the Brent Spar.

The UK Government announced its approval of the Abandonment Programme in February 1995, and also informed the thirteen European governments which are signatories to the Oslo-Paris Conventions. No objections were raised.

Thus all the formal requirements of the system were fulfilled, and all the parties who had rights to input to the process had had the opportunity to do so. However, well publicised action by Greenpeace, discussed in detail in other chapters of this book, forced Shell to abandon their plans in June 1995, while Brent Spar was being towed to its disposal location. The UK Government reluctantly accepted Shell's decision, and helped them to obtain a licence from the Norwegian authorities allowing the Spar to be anchored in Erfjord.

The UK Government subsequently made it clear that the only approved BPEO for the Brent Spar would remain deepwater disposal, and that any alternative Abandonment Programme would have to be demonstrably better than deep-sea disposal. Shell is now engaged in a lengthy consultative process aimed at finding a solution which will match or better deepwater disposal.

The Brent Spar experience shows how public opinion can be conditioned by environmentally moral arguments, regardless of how well based in fact these arguments are. Shell was not able to react appropriately to this.

Shell have subsequently made the following statements:

1. *"We have certainly learned that sound science is not a sufficient precondition to progress, but that ... we have to address additional elements as well" (Heinz Rothermund [3, p. 125])*

2. *"We are committed to consulting more widely and at each step of our search for a solution this time around. Organising this will be difficult, but our aim is for our final proposal to the UK government to be informed by the views of many other interested and representative parties. In the end, however, it will be up to Shell alone to complete the BPEO analysis and arrive at a balanced solution to recommend for the Government's approval. If the exceptional events surrounding Brent Spar have achieved a good outcome, we hope it has been to place such difficult environmental choices and the contribution science can make to finding solutions at the forefront of many more minds. We will continue to defend the balanced approach to environmental decisions in the belief that it is in the vital interest of our economies, our societies and the environment. But we will also pay much more attention to listening to and consulting people about the many issues involved and to gaining their confidence and trust." Shell spokesperson [11]*

3. *"With the benefit of hindsight I see two main lessons for companies from the Brent Spar affair.*

   The first is that we were slow to appreciate that the main focus of the Greenpeace campaign was directed not at British audiences but at those in other countries. This demonstrates the changing pressures on businesses in today's 'global village'. Learning to communicate across borders - to those with different cultural preconceptions and less understanding of the local context - will be a growing challenge for all multinational companies.

   The second is that the debate about how to reconcile wealth creation with environmental protection - how to achieve sustainable development - is not a nine-week affair but a continuing process of profound relevance to all people. It is vital that the business voice is heard. This requires companies to be more open about their activities, more ready to debate the issues they raise, more responsive to people's concerns, and more assiduous in demonstrating that they can be trusted to exercise their power responsibly." Chris Fay, Shell UK's Chairman and CEO [15]:

4. *"We have accepted that a key implication of the Brent Spar affair as we look forward is that technocratic compliance with rational, science-based regulation is not, in itself, enough. Shell and the offshore industry must engage successfully in a very much wider political and public domain where the trust, confidence and 'licence to operate' from the public at large has to be won.*

   The Shell argument ... was based on science, reason, and balance; it needed time, and a degree of patience with scientific arguments and data to be successfully explained and understood. We now readily admit that we did not anticipate this early enough. Nor did we anticipate the unprecedented extent to which widespread emotions and fears can be aroused when a carefully balanced, scientific argument finds itself under attack from an aggressive "single issue" campaign, freely exploiting dramatic visual stunts and colourful, often inaccurate soundbites.

   Brent Spar became a clash between the head and the heart; a conflict in which scientific reason and systematic, balanced judgements were set against the power of emotion, fear, and even myth. Particularly in northern Europe, the Brent Spar became a symbol of man's misuse of the oceans, irrespective of the reality. The campaign exploited images of grand scale 'litter-louting', of David and Goliath, of heroism and derring-do. And, alarmingly, the arousal of such emotions also brought escalating violence and echoes of anarchy.

   ... two important considerations led to the conclusion that the company's position had become untenable. One was a radically altering balance of the safety risks which had been, paradoxically, a key element in the disposal decision at the outset. Also, with European government ministers now openly opposing the plan - not on its rational merits, but on its symbolic significance - Shell was faced with a seachange to the established international framework of regulatory principles and standard which underpins our 'licence to operate'.

   We have learned that it isn't enough to be right. We accept that we should have consulted more widely and earlier. We believe, too, that businesses such as ours must now find new ways of absorbing not only the rational arguments of the best-informed, but also widely held, deep-seated emotions and concerns. We must learn to understand and become more adept at addressing the power of symbolic gestures, and we must work harder at encouraging people to think about the trade-offs between the benefits industry brings, and the ultimate costs of these - both economic and environmental.

We have learned that we must come better to grips with the way that emotional and ideological concerns can resonate with politics. We must also recognise better the cross-border dimension of issues in the new Europe, where emotional outrage and consumer action in countries outside the UK can lead to the reversal of a plan for a UK asset which complied fully with UK and international regulation.

It is clear that there is a need for wider consultation and debate on the issues surrounding not only the decommissioning of offshore oil installations, but many other matters like it. Reason and science must remain the basis for environmental decisions, but they are not in themselves enough. Shell UK and the rest of the UK offshore oil industry are now engaged in seeking to involve people more in thinking about how the benefits of industry on the one hand can be balanced with managing environmental, safety and health risks on the other. We certainly accept that we must communicate openly and widely, and listen better to people's concerns. We look forward to joining an effort which involves scientists, environmentalists, and policy-makers, and, by whatever means best present themselves, the widest possible range of all our public stakeholders.

Difficult decisions will stand or fall on the extent to which they command public support and widespread understanding. We all depend on rational consideration of the facts, and we depend on our elected politicians to balance varied and complex considerations on behalf of the whole community. Single-issue campaigners can play a valuable role in arguing for a particular view, but alarmist exaggeration helps no-one. The intrusion of the firebomb and the bullet into the Brent Spar debate was a profoundly worrying development." John Wybrew, Shell UK's Director of Corporate Affairs [16]

The Future

Figure 1 shows an estimate (using cost estimates and approximate timings from Wood Mackenzie [14]) of the number of structures to be decommissioned, by year, and their approximate costs. For example, in 2011 it predicts 9 structures to be decommissioned, the largest costing £1000M, the smallest £15M, total expenditure for that year being £2300M. The significance is not in the absolute values, which are necessarily speculative, but in the relative magnitude through time. The shape of the plot illustrates how poorly our experience equips us for the future.

Significant technical problems will have to be faced first with the decommissioning of large steel platforms in deeper water. For example NW Hutton, Heather, Maureen, Murchison, Beatrice, Kittiwake, Hutton and Thistle could all face decommissioning in the next five years. The sums involved in

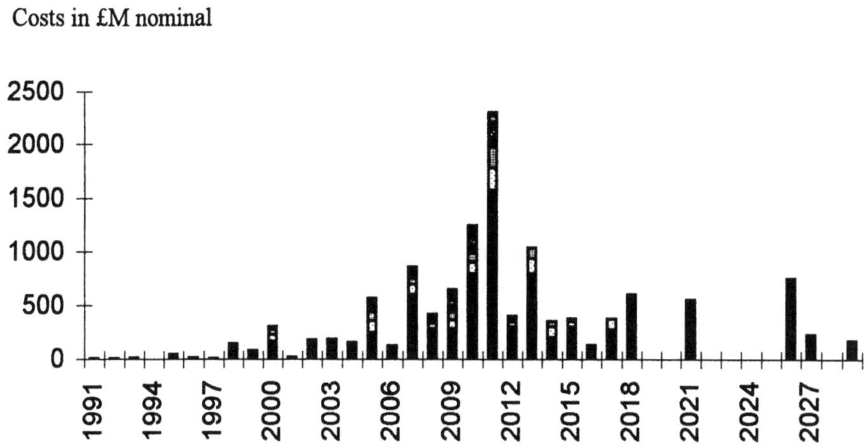

Figure 1. Estimated costs of decommissioning

decommissioning each of these large structures could be as much as £100M [14]. Even greater problems will be faced with regard to the large concrete gravity-based platforms, the first few of which to be decommissioned are likely to be those at Ninian, Dunlin, Cormorant and Brent, between 2007 and 2011. These projects are way beyond current industry experience and will present considerable problems with regard to safeguarding the environment. It is not at all clear, technically, whether these structures can be moved at all, without some form of structural failure, primarily because of the suction effect beneath their bases. Indeed Greenpeace [3 p.72] acknowledge that it may well be a better solution, from an environmental point of view, to leave the bases of these structures in place. The feasibility of removing and taking ashore the topsides for disposal is well established, as Shell has shown in the case of their redundant modules on the Brent Field platforms.

From an environmental point of view, one of the more serious issues may turn out to be that of cuttings piles, which, as discussed earlier, can be large and can contain significant volumes of hydrocarbons. Most have been on the sea-floor for some time and have stabilised, some forming "crusts" which seem to retain the hydrocarbons. It is clear that in the process of removing platforms, these piles will be disturbed and their hydrocarbons released into the marine environment. The industry is currently studying ways of dealing with cuttings piles, including "hoovering" them up, or finding ways of decommissioning without disturbing them. At this point there is no clear preferred solution.

A major problem is how to portray to the public the full spectrum of views, in a balanced way. Gray [3, p 185] points out, for example, that "probably the most intrusive disturbance to the ecosystem of the North Sea is commercial fishing and especially trawling ... it has been estimated that every square metre of the North Sea is trawled over twice, and some areas over six times, per year." But how often does this side of the argument appear in the media?

UKCS Operators, through UKOOA, remain committed to doing all that is reasonably practicable to reduce the safety risks and the risk of damage to the environment when decommissioning offshore installations. To this end they are developing efficient, fit-for-purpose decommissioning solutions, based upon achieving the right balance between safety, protection of the environment and cost, for those installations which have reached the end of their economic life. It is to be hoped that the lessons learned from all of industry's experiences, and particularly those of Brent Spar, will allow the management of the decommissioning process to proceed smoothly, achieving minimum environmental impact with maximum efficiency.

References

1. North Sea Infrastructure, in "The North Sea - A Province of Excellence", pp 84-89, a Petroleum Economist publication, December 1995.
2. M. Quinlan, Troll breaks all records, in "The North Sea - A Province of Excellence", pp 84-89, a Petroleum Economist publication, December 1995.
3. House of Lords Select Committee on Science and Technology: Decommissioning of Oil and Gas Installations, Session 1995-6, HL Paper 46-I.
4. Geneva Convention on the Continental Shelf, 1958, Art 5(5).
5. United Nations Law of the Sea Convention (UNLOSC), 1982.
6. IMO (1989). Guidelines and Standards for the removal of offshore installations and structures on the continental shelf and in the exclusive economic core. Resolution A, 672(16) of the International Maritime Organisation, October 1989.
7. DTI Oil and Gas Office, Aberdeen, "Consultative Document, Guidance Notes for Industry, Abandonment of Offshore Installations and Pipelines under the Petroleum Act 1989", May 1995
8. AURIS Ltd, August 1995, "An Assessment of the Environmental Impacts of Decommissioning Options for Oil and Gas Installations in the North Sea."
9. UKOOA, 1997. World Wide Web Page, http://www.okooa.co.uk/
10. Department of Trade and Industry, 1997. The Energy Report, Oil and Gas Resources of the United Kingdom, vol. 2.
11. Shell, 1997. World Wide Web Page, http://shellexpro.brentspar.com.
12. Rudall Blanchard Associates Ltd 1994. Brent Spar Abandonment BPEO. Prepared for Shell UK Exploration and Production
13. Rudall Blanchard Associates Ltd 1994. Brent Spar Abandonment Impact Hypothesis. Prepared for Shell UK Exploration and Production.

14. Wood Mackenzie North Sea Service, Upstream Oil & Gas, UKCS Summary Field Data, Fields in Production/Under Development, June 1996.

15. C. Fay, 1996. Not Black and White but Shades of Green - a Business Perspective on Reporting the Environment, a speech delivered to a media conference at the University of Wales, 20 May 1996.

16. J. Wybrew, 1996. Brent Spar - the Implications for Environmental Decision-Making and Public Support, an article written for the "Marketing Managers' Yearbook 1996", abstract on Shell's World Wide Web Page, see 11. above.

2

Technical Aspects of Decommissioning Offshore Structures

P A Meenan

Introduction

The options for decommissioning offshore facilities and structures are many and various. They are driven by environmental protection, cost, health and safety, available technology and politics. This chapter examines the range and type of facilities and structures which will be decommissioned; the framework of legal requirements; the onshore sites at which the work can be done; the possibilities of reuse and materiel recovery; the methods available and possibly requiring development for cutting, toppling, lifting and transportation, and the particular problems associated with the larger fixed structures.

The nature of the very large concrete gravity sub-structures (CGSs) and the difficulties associated with their decommissioning is reviewed and discussed in some detail. Economic aspects of decommissioning management are introduced with comparisons to other industries, and some discussion is made of the possibilities for recycling of valuable materials and reuse of plant and equipment.

The intention of the chapter is to advance the debate rather than to offer specific solutions, although some new ideas are presented in relation to the particularly problematic large concrete gravity sub-structures.

The infrastructure of Offshore Oil and Gas Installations which has been created in the North Western European Continental Shelf (NWECS) represents an investment of £150Bn in the period 1965-1996. This cumulative investment has enabled the United Kingdom to produce gas to allow the power generation industry to become much more efficient and to achieve independence from coal production.

For power generation the economics have been shaped by political will which has "surfed" opportunistically on the waves of technological change created by the oil and gas offshore and onshore industries. The switch from area extensive and site specific facilities for coal fired power generation, to small-site universally-locatable (on the gas pipeline and electrical overhead grids) gas fired power stations, has been based on two principal factors: firstly the short term economic value of the large areas of land released from the old coal-fired power stations, and secondly the failure of government to place a realistic and socially responsible burden on the gas-consuming power generation companies to make provision for the whole-life cycle decommissioning costs of the offshore oil and gas production facilities.

Setting the Scene

The United Kingdom Operator's Association (UKOOA) [1] has published a very good and concise historical perspective of the UKCS Oil and Gas Industry from which the following is summarised.

From the first natural gas discovery in 1965 and first oil discovery in 1970 Britain became self-sufficient in oil by 1981. At an age of 31 years the British oil and gas industry is providing 75% of the country's primary energy needs and has contributed approximately £140 Billion in royalties and taxation while making a cumulative capital investment of approximately £150 Billion. At the current level of production the annual contribution to the UK Balance of Payments is approximately £4 Billion. Compared to this, UKOOA estimate that the total cost of decommissioning offshore installations in the UK sector will be £5 billion with £1.5 billion expended in the ten years 1996 to 2006 on the first 50 installations to be decommissioned.

Physical Statistics

By the end of 1996 there will be 140 offshore fields producing oil and gas from over 200 installations in the UK sector. These installations are fixed steel jackets, concrete gravity substructures and floating/subsea production systems. The older installations are mainly fixed steel and concrete structures, generally of a major scale in the northern area and in the southern area mostly smaller scale steel jackets [2].

Approximately 5000 wells have been drilled of which half have been exploration and appraisal and have been presumably abandoned satisfactorily (plugged and cut). The remaining 2500 are development and production wells most of which remain to be dealt with as part of the decommissioning programme for each facility.

It should also be remembered that there are 5000+ miles of pipeline (on and offshore), substantial onshore processing facilities (e.g. Mablethorpe, St Fergus, Mossmorran, Sullom Voe, Flotta), and storage/tanker terminals (Sullom Voe, Flotta, Nigg, Braefoot Bay, Hound Point) which will also need to be decommissioned.

Responsibility and Engineering Leadership

If those who fully understand the problems, the moral and ethical issues of offshore platform decommissioning and the full range of technical complexity do not step into the public debating arena, then what chance is there of a socially, economically and politically acceptable solution? The social sub-group who knows in an objective, calm and rational manner the total extent of the problems and their solutions is the body of professional engineers engaged by the various parties, often on "opposing sides", in the management structures of the operators and the design and installation contractors.

The only way that society can decide successfully the questions of offshore platform decommissioning is via its engineers and applied scientists who truly understand the global environmental and ecological issues and dynamics.

The recent experience of Shell UK with their Brent Spar disposal plans has clearly established a benchmark for those engaged in designing decommissioning/ abandonment projects. Every step of the process must now be communicated to the international public, their understanding gauged and their approval sought. Their approval will of course never be fully measured but quite assuredly their disapproval will have major economic consequences for any operator who fails to communicate his plans [3]. Engineers must, therefore, take the responsibility for explaining the technical issues and decisions to the public pro-actively via the public affairs professionals of the oil companies by whom they are employed and more importantly via their professional institutions .

Decommissioning Sites

The potential decommissioning sites are all the fabrication sites in UK, Norway, Sweden, Denmark, Holland and Germany, at which the facilities were originally constructed (either currently operational or in care and maintenance). Those sites with major graving docks or dry docks are likely to be best placed to handle the very large de-constructed elements such as sections of jackets, module support frames, large modules and integrated decks. These sites which are also located close to a steel making/recycling infrastructure will be much more economic to use due to reduced transport land based transport costs.

Topsides modules, of the first and second generation of platform design (i.e. 500 to 2000 tonnes) can easily be decommissioning and scrapped, or reconditioned for other projects at the smaller fabrication yards.

There is no technical aspect of decommissioning and retro-loading to shore which will require or justify the creation of new sites. Given that there will be no "urgency to first-oil" dynamic applying to the decommissioning process and the consequence that the extra premium for travelling labour will not be affordable, then it is probable that the bulk of the work will go to the sites closest to the industrial centres of population. This may offer a beneficial social opportunity to regenerate industrial activity in the former shipbuilding areas of the country. It is also possible however that the very low cost ports of the former Soviet Union (FSU) and the Warsaw Pact countries could become very significant competitors.

This would be virtually certain if a market emerges or can be created in the FSU oil production theatres for reconditioned North Sea plant.

Politics and Economics

The public relations impact and loss of revenue suffered by Shell over the Brent Spar decommissioning and disposal attempt was a salutary lesson which we have all taken to heart. Despite the rigour and excellence of the Shell Expro Aberdeen Engineering and Environmental studies, and notwithstanding the fact that both the Department of Energy and the Government were in full agreement and support of the deep sea dumping option, the supra-national power of European consumer opinion forced Shell International to attempt to seek another solution which almost certainly will be more expensive, less safe to the employed workforce, will consume a large amount of energy and will entail risks to coastal navigation routes and environments.

It may be that there is no alternative to this situation. It may be that the romantic, ideological, aesthetic (or whatever) desire to keep the mid-ocean abyss free from any man-made artefact—toxic or innocent—is so strong and will remain strong for the foreseeable future, that we will be denied the most efficient "Best Practicable Engineered Option" for some platforms. That of course is Democracy and we must accept it [4].

Legislative and Statutory Requirements

The dominant requirements are those instruments of Navigational, Health and Safety and Environmental Legislation, in particular:

- Oslo and London Conventions, Guidelines 1991
- Petroleum Act 1987
- Offshore Installations (Safety Case) Regulations 1992
- Coast Protection Act 1949
- Environmental Protection Act 1990 (Part II)
- Food and Environment Protection Act 1985
- Radioactive Substances Act 1993
- Offshore Safety Act 1992
- Health and Safety at Work Act 1974

The following Guiding principles are an over simplification but provide a useful overview [5]:

"A case-by-case approach will be maintained."

"An abandonment programme will be required in respect of all offshore installations and pipelines."

"After 1 January 1998 no installation or structure shall be placed in the UKCS unless its design and construction is such that entire removal is feasible."

"Abandonment of an installation or pipeline may be deferred if such action can be justified."

"The persons who own an installation or pipeline at its abandonment will normally remain the owners of any residues. Any residual liability remains with the owners ***in perpetuity****. Owners will also be responsible for complying with any conditions attached to the approval of the abandonment programme."*

The application of the Oslo and London Conventions to the basic options is as shown below:

Disposal Method	**Oslo Convention**	**London Convention**
Leave in situ	Not addressed	Practice confirmed as dumping (1). Special permit required (using DTI legislation). Issue of the permit must be notified to the LC'72 Secretariat. (2)
Topple at site	Case for dumping rests on contaminants present (Annexes I and II), the case that no serious obstacle to navigation would result (Annex II, I(b), 2 and 4 and the general provisions of Annex III.	As above
Move to side and dispose	As above	As above
Remove and sink in shallow water location	As above	As above
Remove and sink in deep water location	As above, but if dump site is at least 150 nautical miles from shore and depth not less than 2000m, no need to consider 'interference' issues.	General permit required (as no serious obstacle to navigation or fishing would result) subject to satisfying 'contamination' criteria.

Table 1. Oslo and London conventions [5]

In September 1992 contracting parties to the Oslo and Paris conventions adopted the new OSPAR Convention. Annex III of this convention deals with the prevention and elimination of pollution from offshore sources and includes specific reference to the dumping of disused offshore installations and pipelines. The OSPAR Convention has not yet been ratified but when in force will replace the existing Oslo Convention.

Decommissioning Options

The best options inevitably appear to be different to different special-interest groups. This is of course not surprising. The principal spheres of special interest are the Environmental, Health and Safety, Financially Interested and Politically Active groups. The literature and the established conventional wisdom has identified the Best Practicable Environmental Option as the paradigm for accepting an abandonment strategy, but it is necessary to examine what is Best Practicable from each of these spheres of interest. In this chapter it is proposed that these options be defined :

- Best Practicable Environmental Option **BPEO**
- Best Practicable Safety Option **BPSO**
- Best Practicable Financial Option **BPFO**
- Best Practicable Political Option **BPPO**

An integrated definition of "Best" is required. The unification of all these options is perhaps usefully termed the Best Practicable Engineered Option (**BPEngO**).

Within each sphere of option review there is a common set of physical actions or sequences of action:

1. Do Nothing (i.e. except navigational lights and buoys)
2. Remove Topsides to shore
3. Remove Topsides and Part of Substructure
4. Remove Topsides and Whole Substructure
5. Remove Topsides, whole substructure and cut Wells and Piles to -5metres

The "best" action priorities are likely to vary radically across these groups. It is likely that the Environmentalists will want to get to stage 5 as soon as possible and will not be satisfied with anything less in the long term but may, if reasonable, be prepared to accept stage 2 for the short term. Politicians, either in power or in opposition, are likely to pursue the short term achievement of stage 5 in the interests of job creation and winning green votes. Beyond the medium term (i.e. the next general election) they are probably interested in only the minimum action.

D = years from last oil/gas	**BPEO**	**BPSO**	**BPFP**	**BPPO**
Short Term 0 < D < 3	2	1	1	5
Medium Term 3 < D < 5	¾	2	1	3
Long Term 5 < D < 10	5	3	2/3	1

Table 2. Differentiated "BEST" action priorities

- Physical Options—fixed, floating and sub-sea facilities
- How do we reconcile the Best Practicable Engineered Option to meet all the Environmental, Safety, Financial and Political requirements?
- All structures do not have the same range of options
- Are we prepared to recognise that for some platforms the BPEngO ***may*** be to do nothing short term and only topside removal long term?
- The BPEngO will be driven by creative management operating on the international supply chain, incentivised by innovative fiscal measures by Government.
- Options for Decommissioning/Removal must be reached on the basis of a coordinated and mutually agreed approach for all NWECS sectors (particularly UK and Norway).
- The Deep Dumping (Abyssal Emplacement) Option is too valuable for disposal of certain innocuous major elements to be neglected.
- We must not be afraid to persuade international public opinion of this and we must explain why.

Technical Issues

Scope of Work

The offshore decommissioning scope of work which will have to be undertaken will relate to four basic classes of installation i.e. Fixed Platforms, Moored or Tethered Platforms, Pipelines and Subsea Structures. Of these the fixed platforms (i.e. jackets and CGS structures) represent the greatest difficulty and engineering

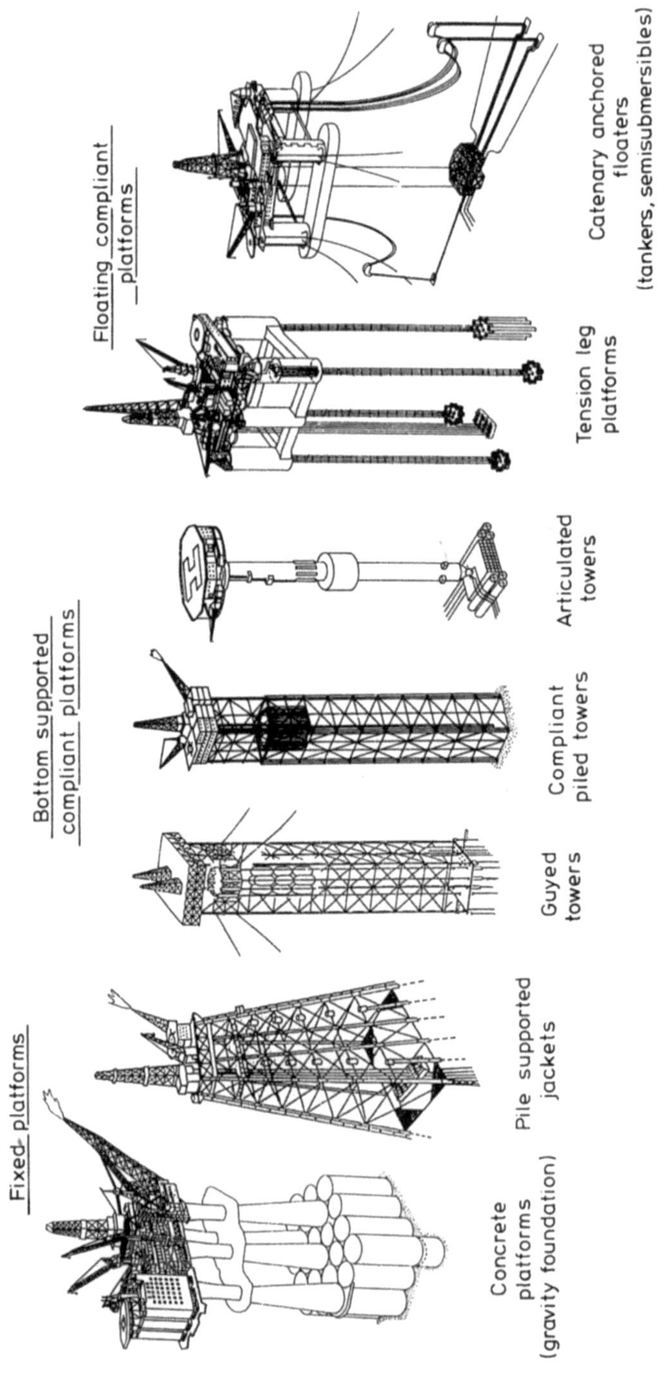

Figure 1. Offshore structures for exploration and production (reproduced from G. Clauss et al: Offshore Structures Volume I, © Springer-Verlag 1988)

challenge, while the moored/tethered facilities will be relatively easy to remove due to the inherent nature of their design (Fig. 1).

Fixed Structures

The legal options are defined in the following table.

Installation	Water Depth *	Weight (tonnes)	Complete Removal	Partial Removal	Toppling	Leaving in Place
fixed steel	<75m	<4000	YES	No	No	No
fixed steel	>75m	>4000	YES	Yes (1)	Yes (1)	No
fixed concrete	<75m	<4000	YES	No	No	No
fixed concrete	>75m	>4000	YES	Yes (1)	Yes (1)	Yes (2)
floating	<75m	<4000	YES	No	No	No
floating	>75m	>4000	YES	No	No	No
column (eg ALC)	<75m	<4000	YES	No	No	No
column (eg ALC)	>75m	>4000	YES	Yes (1)(3)	Yes (1)(3)	No (1)(3)
sub-sea	<75m	<4000	YES	Yes (1)(4)	Yes (1)(4)	Yes (1)(4)
sub-sea	>75m	>4000	YES	Yes (1)(4)	Yes (1)(4)	Yes (1)(4)

Table 3. Regal options [5]

Notes

1. Minimum water clearance of 55m is required above any partially removed installation which does not project above the surface of the sea.
2. Would be a measure of last resort and apply to support structure only.
3. Column removed entirely; anchor arrangements base would be addressed separately.
4. Would depend on dimensions, configuration, type and east of removal of item(s).

*100m for installation emplaced on or after 1/1/98.

The options illustrated in the table above should be regarded as *indicative*. The option selected in a particular instance will depend on the outcome of a comparative assessment (Fig. 2).

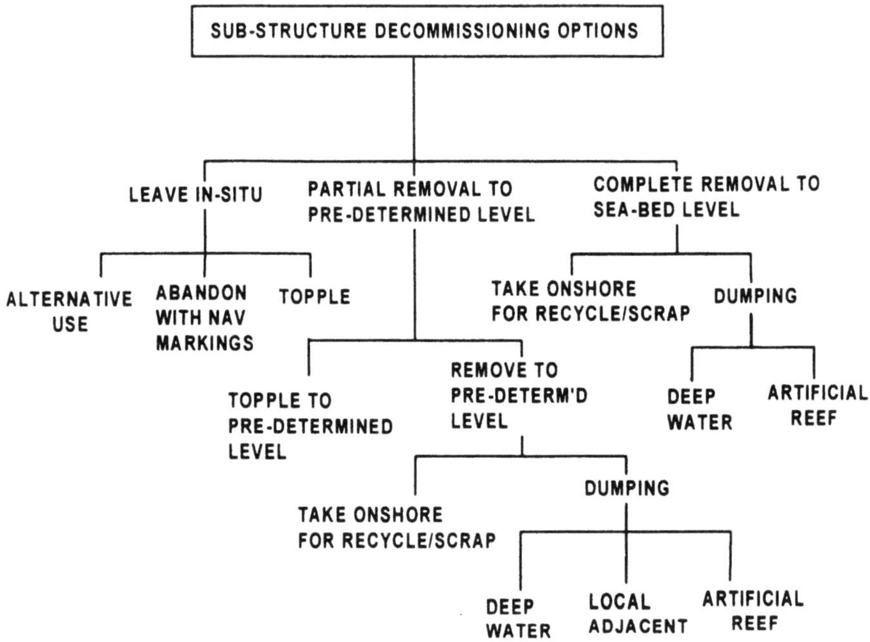

Figure 2. Decommissioning options

Topsides

The topsides of fixed structures are either modular packages of 500 to 3000 tonnes dry weight (i.e. without operating inventories) stacked on plate girder Module Support Frames (MSFs), or Super Modular Packages (SMP) of 3000 to 5000 placed on top of Integrated Deck Structures (IDs) up to 9500 tonnes. The former configuration is to be found on the older platforms and the latter on the newer platforms installed after the availability of the very large semi-submersible crane vessels (VLSSCVs) (ca 1988). The typical weights of topsides, MSFs, IDs and modules are shown in Appendix: Table A1 [3].

One option which has been discussed by a number of operators and which is acceptable under the legislation in some circumstances, is to clean up certain modules to be lifted and placed on the sea bed immediately adjacent to a partially removed or toppled jacket. Given the mobilisation and operating costs of the necessary heavy lifting operation it is obvious that the additional cost of placing on a transportation barge or on the deck of the Semi-Submersible Crane Vessels

(SSCV) for retro-transport to an onshore decommissioning site would be marginal. It is difficult to see a convincing argument for sea-bed emplacement of any component of topsides equipment or structure anywhere on the continental shelf.

Concrete Gravity Sub-structures (CGS)

Platforms of this type (sometimes described as GBS—Gravity Based Structures (Fig. 3)) are far fewer in number but are one or two orders of magnitude heavier than their depth-equivalent steel jacket neighbours. They present particularly difficult problems for decommissioning and removal. Removal of topsides modules can be achieved by SSCV or VLSSCV as for steel jacket platforms.

The Module Support Frame (MSF) incorporated in these structures however, takes the form of a major bridging deck steel structure—usually of box girder or space frame construction. These decks were installed on the concrete shafts by floating them over on support barges, with the concrete sub-structure submerged to low-freeboard at a deep water inshore site. They are very heavy because of integration of process and utilities plant. Generally they are beyond the lift capacity of even the largest VLSSCV (as a single unit).

The conventional wisdom which has prevailed without re-examination until the Brent Spar situation, has been that the existing Concrete Gravity Sub-structures on the United Kingdom Continental Shelf (UKCS) will be removed at abandonment by refloating and dumping in mid-Atlantic abyssal trenches. There are questions to be answered about the technical practicality of de-ballasting, disengaging from soil penetration and structural integrity during these processes [6; 7; 8]. These questions relate to the very large first generation of CGS designs (Fig. 4). For later and smaller CGS concepts, such as Ravenspurn North, NAM F3, and Unocal Q1 and also for the latest CGS - BP Harding - the requirements for removal by de-ballasting have been carefully addressed by the design foundation concepts and in detailed design of non-corrodible pipework systems (for example) necessary to ensure this.

It appears, however, that very little thinking or study work has been applied to the large first-generation concrete gravity substructures (CGS) or gravity based structures (GBS) as they are sometimes labelled [8]. Perhaps the enormity (literally) of the problem of removal has been seen as too daunting to tackle until the time is reached when these installations are no longer producing economically. Given that oil price is capable of dropping and that the predicted depletion curves for the reservoirs could be erratic in the post-mature phase, it is possible that technical solutions together with their safety and environmental cases (without which there are no solutions) may be required earlier than expected.

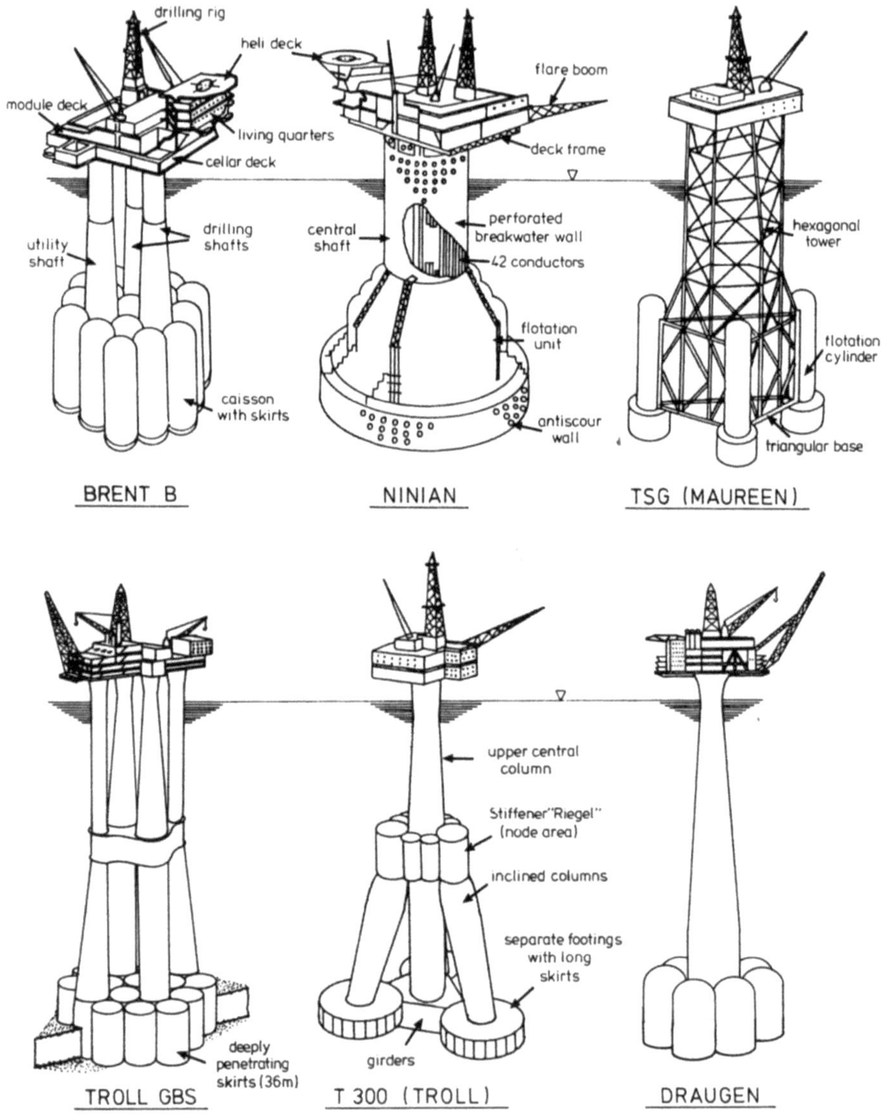

Figure 3. Gravity-type concrete and steel platforms (reproduced from G. Clauss et al: Offshore Structures Volume I, © Springer-Verlag 1988)

Because deep water dumping (or abyssal emplacement) in the mid-Atlantic trench is not now, and may not ever be, acceptable, the time has come to address technically and economically the methods which may be developed for partial removal of the upper zones of the large CGS platforms.

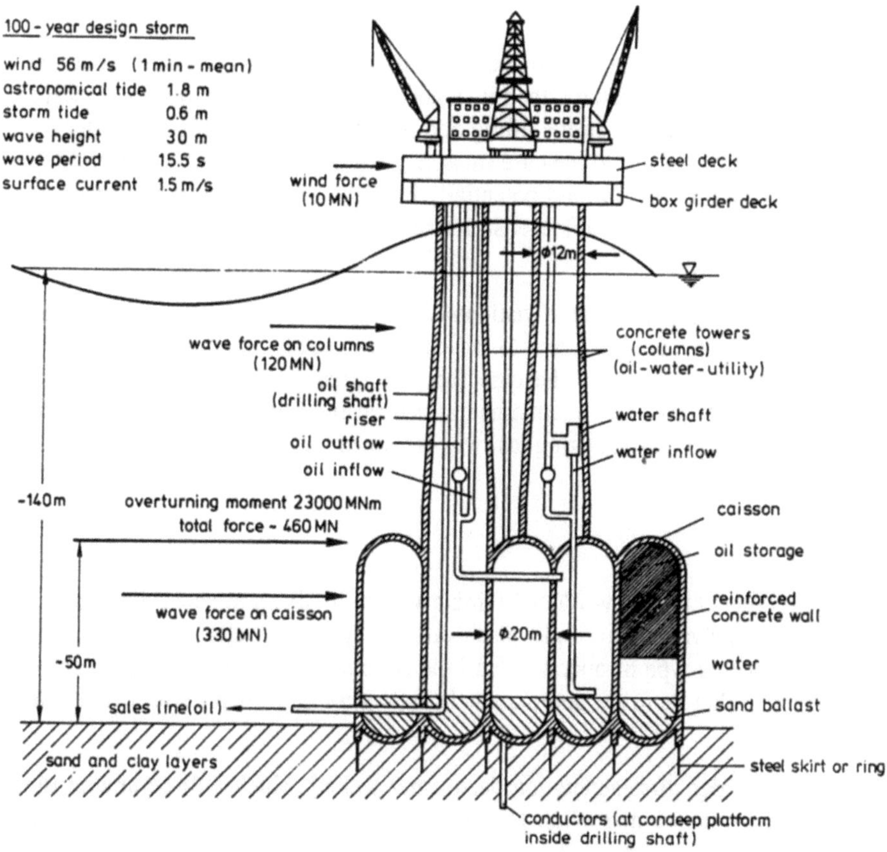

Figure 4. Details of a gravity-type concrete platform (reproduced from G. Clauss et al: Offshore Structures Volume I, © Springer-Verlag 1988)

It may be that the Best Practicable Engineered Option will be to maintain the integrity of the base caissons and to cut and remove the shafts from the top of the caissons.

The greatest environmental risk is the possible release of residual oil, slops and sludge from the storage cells should the upper domes or slab be breached by impact during partial demolition or toppling of the shaft using explosives. It is possible that even after purging and surfactant flushing by means of the existing pipework systems a typical oil storage CGS base caisson may be left with a significant volume of trapped residual hydrocarbons. Even if the residuals could be fully cleaned out (or if their release to sea could be permitted for later surface skimming within a boomed area) the removal of the base caissons by progressive underwater demolition would require a cumulative explosive energy which could have a significant impact on fish over a wide area and would entail a high risk underwater operation.

If it can be justifiably argued and accepted that the intact base caisson should be left on the seabed in perpetuity, then it should also be acceptable to dispose of the concrete shafts by cutting, heavy lifting and placing them horizontally on the sea bed immediately adjacent to the base caisson (as they would have a lower profile). In this configuration they would potentially have the same value as fish breeding reefs formed from abandoned jacket structures. (If this is a true value in the North Sea ecology—see [4].)

It is worth noting that even if national and international public opinion and political attitudes should come to accept deep water dumping (or abyssal emplacement) it is probably the case that the large CGS platforms could not practically, safely and with maintained structural integrity, be removed by base pressurisation and de-ballasting as was assumed (cursorily) by the original designers.

Jackets

Jacket structures in shallow water having a weight in air up to about 5000 tonnes can be lift removed in a single piece (after cutting of wells and piles). The economics of this type of operation and the commercial risk would appear to be feasible given that there are approximately eight vessels available worldwide with the capacity.

Some of the older, first generation deep-water jackets which will be among the first to be decommissioned weigh 15000-20000 tonnes and will need to be cut and partially removed or partially cut and toppled. The requirement for these substructures is to achieve clearance to -55 metres (LAT) and because the weight distribution of these older jackets is light at the top and very heavy at the bottom, it may be that the top section above the -55m cut line will be less than 5000 tonnes and therefore removable as a heavy crane lift. The corollary of this is that a VLSSCV based operation to achieve -55m removal may be able to achieve greater clearance (maybe -75m) with no significantly greater cost for a single stage cut/remove cycle. Removal of a major jacket in its entirety could be achieved by 3 or 4 heavy lifts.

There are significant engineering problems to be overcome and temporary works/lifting and support appurtenances to be designed, fabricated and installed. For example the "stiction" (suction, adhesion, embedment and friction) of the mudmats will have to be eliminated by water injection and jetting before the final bottom sections could be lifted. It should also be clearly understood that the precision of very heavy lifting operations, required by safety procedures and crane vessel operating limits, means that lifting attachment points, load path determination and component stress analysis must be fully engineered. The operations will not remotely resemble the "grab and tear" character usually associated with industrial demolition or ship-breaking.

Moored or Tethered Structures

This group of structures consist of Tension Leg Platforms (TLP), Floating Production Systems (FPS), Floating Production Storage and Offtake (FPSO) systems, Spar Buoys and Articulated Towers. All of these are vessel-based and therefore have buoyancy. Operational safety and maintenance requirements result in all mooring lines and tethers being relatively easily disconnectable both from the vessels and from sea bed anchor points. Flexible flow lines and risers are similarly disconnectable and retrievable.

There is no need to discuss further the decommissioning aspects of this type of facility, other than perhaps to point out in passing that the very large FPSO vessels which are emerging on current projects will only be able to access a relatively few sites for decommissioning or recommissioning for other fields. (i.e. those sites with major dry dock or graving dock facilities).

Several concept designs have been produced for concrete FPSO hulls (e.g. Schiehallion) and off West Africa there is one large concrete FPSO hull (Nkossa). The draught of such designs which might be realised in future will almost certainly require decommissioning at a near shore mooring rather than alongside a quay or in a dock.

Pipelines

The extent and nature of the offshore pipeline infrastructure on the UKCS is shown in Appendix Table A2.

The technical options are

1. Remove to surface, then shore.
2. Bury
   2.1 by retrenching
   2.2 by rock dumping
3. Part Remove
4. Leave in Place

It has been suggested [9] that pipelines could be removed by a reverse lay process using semi-submersible lay barges or by sea bed cutting and lift removal in appropriate segment lengths. It is suggested that reverse lay barge recovery is likely to be more cost effective. No real study has been made of these options but work is now necessary.

A number of offshore installation contractors are of the opinion that sea-bed cutting and segment lifting by smaller derrick barges may be a highly productive less costly operation. This would certainly be true if remotely operated pipeline crawler cutters were developed which could be deployed in advance from a Diving Support Vessel (DSV) or perhaps a less expensive supply vessel or anchor handler.

Where pipelines have been successfully buried, either by trenching or by covering, and have remained buried there would appear to be very little point in attempting removal. Where such pipelines have been locally uncovered by sea bed sediment transport they should probably be removed as the long term stability of replaced material could not be assured.

Sub-Sea Structures (Medium to Deep Water)

These structures range from guides bases for exploration wells which have been left in place on top of an uncut conductor pile, through single wellheads, to multiple wellhead templates and production manifolds.

In shallow to medium water depths, the removal of these elements can be achieved entirely with existing cutting technology and relatively small lifting vessels (i.e. derrick barges or even deck cranes on SS support vessels. In deep

ITEM	ABANDONMENT TREATMENT
Drilling template (platform)	Should be regarded as part of the jacket structure.
Drilling template (satellite)	Should be entirely removed with any piling cut at a suitable depth.
Production manifold (small to medium)	Should be entirely removed with any piling cut at a suitable depth.
Production manifold (large)	May require special consideration: removal of fittings should be undertaken.
Xmas trees	Should be entirely removed.
Well heads	Removed as part of well abandonment.
Protective structures	Should be entirely removed unless there is some benefit in retention.
Anchor blocks (concrete)	Should be entirely removed if small: larger blocks would require special consideration.
Anchor points (piled)	Should be entirely removed with any piling cut at a suitable depth.
Riser basis (piled)	Should be entirely removed with any piling cut at a suitable depth.
Anchor chains/cables	Should be entirely removed.
Risers (rigid and flexible)	Should be entirely removed.

Table 4. Regulatory framework for abandonment treatments [5]

water (i.e. beyond direct diver intervention) such as West of Shetlands the design of such sub-sea units necessarily means that the installation process is mechanically reversible under remote operation by specialised equipment such as BPs DMAC (Diverless Maintained Cluster) attachment units. Only if such equipment or procedures fail irreversibly should it be necessary to seek approval to abandon the equipment on the sea bed, even though in deep water the required 55m LAT clearance will be achieved everywhere without removal.

The options for abandonment treatment of sub-sea structures and equipment, is to comply with the regulatory framework, as shown in Table 4.

Modification for Reuse

The prospects for refit, reconditioning, or re-engineering of FPS and FPSO vessels would appear to be very sound. It should be borne in mind, however, that the characteristics of oil and gas producing fields and their reservoirs can differ over a wide range of parameters such as gas/oil ratio, specific gravity, sour gas content, water cut, sand content, pressure and temperature. This means that unless an operator has, or knows of, an upcoming field development whose parameters match approximately those of a facility which is about to be, or has just been decommissioned then it is very unlikely that a reuse could be phased successfully given the lead time for re-engineering and refit.

Subsea components such as wellheads and production manifolds may also be reusable especially where they have been designed to a high specification with corrosion resistant materials and deployed for relatively short production life e.g. less than 10 years. For very deep water sub-sea equipment, such as the Foinaven wellheads and production manifolds which have been engineered and fabricated to very high standards, and which will not experience external corrosion over a short production life, the prospects for reuse would also appear to be good. The all-up costs of such deep sub-sea components and structures can be around £50,000 per tonne.

Reuse of fixed major platform structures (i.e. other than topsides equipment) is almost certainly not feasible. Steel jackets will be well through their original design fatigue lives and will have been subject to significant thickness loss due to corrosion. They will not be recoverable whole. It is possible however that certain high value elements of the later generation of structures such as cast-steel nodes could be re-examined and found to be fit for reuse.

Concrete Gravity Sub-structures (apart from certain smaller units) will not be removable and will only be reusable in-situ.

Rigs to Reefs

This is an established policy and practice in the Gulf of Mexico [10]. Operators are encouraged to tow their decommissioned steel jackets to designated sites where they are concentrated to provide a habitat and breeding ground for fish and other marine fauna. These sites are popular fishing grounds for recreational SCUBA divers who report prolific fish populations. The policy is on-going and is perceived to be successful.

Given the climatological and oceanographic differences between the North Sea and the Gulf of Mexico, it is very unlikely that "rigs" can be turned into "reefs" with any beneficial effect on fish stocks or distribution. It is probable that there would be no overall increase in the total biomass or the balance of fish types [11]. It is possible that the same total quantities of fish would simply be distributed differently and of course large shoals of fish in, or in close proximity to, such reefs would be of no use whatsoever to standard North Sea fishing techniques and gear.

Recreational SCUBA fishing of the type enjoyed by a large number of people from the US Gulf Coast, is not relevant to the North Sea when one contemplates the locations and depths at which 'reef' sites might be designated.

The recent report of the Natural Environmental Research Council Scientific Group on Decommissioning of Offshore Structures [4] does not offer any encouragement to the rigs-to-reefs concept.

Fish Farming

Coastal farming of fish, salmon in particular, is beset with microbiological problems of infestation and disease control due partly to limited circulation of water in restricted coastal inlets and lochs.

It has been suggested that high value fish such as salmon or halibut might thrive more productivity at an offshore location.

The problems of introducing fish cages adjacent to or within the structure of a jacket type substructure are not inconsiderable. Firstly, there is the problem of man access to water level. This is virtually impossible on safety grounds alone (sea state is 1m significant wave height for 90% of time). Secondly, it would not be possible to have lifting cages within the structure because of geometry. Thirdly, storm damage due to snagging and fretting of nets or cages is likely to be frequent and not easily repaired (or rapidly detected) and finally the cost of logistical support to the fish farm personnel is likely to exceed by a great margin the costs of complex anti-biotic based inshore fish farming regimes.

Wind/Wave Power Generation

Assuming that the two classical and fundamental problems, of electrical power generation are solvable i.e. off-peak storage and cable transmission losses, there are some direct structural issues to be addressed. For wind power generation, to eliminate group loss effects the individual turbine fans, which have to be large in diameter, need considerable lateral spacing (as typified by onshore wind farms). The plan areas of virtually all offshore fixed structures would permit only two to four typical fan turbines. This might produce a total power output of 10 to 20 MW which is unlikely to be economic because of transmission plant costs. Access for intervention and routine maintenance would be high due to logistics support and the need to maintain deck cranes operational.

Wave power generation would require the introduction of a large floating mass to extract reciprocating kinetic energy from transient waves. It is difficult to see how this could be done without entraining considerable more hydrodynamic force onto the modified structure than was allowed in the original design. Strengthening at depth will almost certainly be impossible. Even if fixed platforms were used a mooring structures for tethered buoyant wave power machines, the same problem of increased forces will occur. Added to this is the operational risk of accidental damage, possible failure and consequential aggravated environmental impact.

Military

Whether there could be any strategic defence use for a few specifically located platforms is worth some consideration. It is doubtful if a naval use could be imagined, given that vessel interaction is comparatively difficult. The era of acute concern over soviet submarine movements is over and in any case the platforms are too far south for any useful surveillance and monitoring role.

There may be merit in a very few fixed platforms being retained, with most of the process topsides removed, as air/sea rescue helicopter bases.

The Royal Air Force has from time to time considered the need for a precision 3D radar system located offshore to allow complex real time analysis of aerial tactics in a large volume of airspace. Perhaps a base using four or five suitably located fixed structures might be useful and the strategic economic value might justify the continued maintenance of the platforms.

Navigation/Meteo/Oceanography

Navigation and meteorology is now so developed in the field of global remote sensing and data acquisition by means of geostationary and orbiting satellites that it seems very doubtful if there could be any value in having mid-ocean fixed stations as points of reference or data source.

Oceanographic research in relation to the North Sea might benefit from having one or two platforms for ongoing monitoring of the water column.

Plant and Materials Recovery

This section has been abridged to include sub-headings only in order to keep the length of the chapter to reasonable proportions. This aspect is extremely important and is worthy of detailed discussion on ways and means. No review of the technical aspects of decommissioning has real value if the economic and commercially feasible aspects are not considered in parallel.

The following items will have to be addressed as part of any decommissioning strategy and programme and it is hoped that the other contributors to the symposium will explore possibilities in this direction.

Hazardous Waste

- LSA Scale
- Heavy Metal Sludge
- PCB Fluids
- Halon Gases

Valuable Metals

- Titanium
- Stainless Steel
- Cunifer
- Monel
- Copper (cable)

Reconditionable Plant

- Prime Movers
- Rotating Equipment
- Injection Pumps
- Compressors
- Gas Turbines
- Alternators
- MV/HV Transformers

Bulk Steel

- Predicted tonnages

Economics

- Reverse 'supply' chain management
- Phasing to generate cash flow (or reduce negative cash flow)
- Target markets for recycled material
- Comparisons with other industries e.g. ship-breaking, petro-chemical, nuclear
- Significance of managed safety, preventative maintenance regimes on end of life component values
- Possible fiscal policies, e.g. "subtracted value tax refund"

Cutting Techniques

Study of the options available for decommissioning of offshore structures must necessarily include a review of the techniques, technologies and equipment for cutting of various materials and section geometries. Cutting of the topsides components in the air zone is conventional i.e. it involves only those methods which are known and regularly used for dismantling onshore industrial facilities. Cutting sub-sea is more difficult and although a number of techniques are well proven, there may need to be some development and transfer of technology to enable the scale and volume of cutting to be executed productively.

Cutting of elements which penetrate the seabed and which require to be cut at some depth in the sub-soil requires special techniques such as internal shaped-charge explosive cutting [6], internal high-pressure water and grit erosion [6] and external overshot boring mandrel with internal diamond wire cartridge cutters.

Topsides/Air Zone	Sub-sea	Sub-soil
Oxy-Fuel Gas	Oxy-arc	Diamond wire
Stihl saw	Thermic lance	HP water erosion
Thermic lance	Diamond saw	Shaped charge
Hydraulic shears	Diamond wire rope	
	Shaped charge	
	High pressure water erosion	

Table 5. Applicability of cutting techniques

It is not necessary to discuss the more conventional techniques, but some aspects of the more esoteric methods need to be examined in relation to sub-sea application. Thermic lance cutting is difficult to control in air. It produces large volumes of fumes and because of the rate of burning needs to be held by the operator at a distance of two or three metres, and so is not a very precise method and underwater likely to be difficult to control safely. Its advantage of course is a high rate of penetration and the fact that it can cut cement and concrete.

Diamond sawing is precise and controllable provided that positional control and support is provided in the form of jigs, tracks and attachment fixtures. The depth of cut for the maximum size of saw blades is approximately 600 mm. This could perhaps be extended to 1000mm for use underwater because of the cooling and dynamic damping effect of the surrounding water.

Diamond Wire Rope

Diamond Wire Rope cutting is a well-proven rugged technique originally developed for the quarrying industry and used for a number of years by the oil and gas industry, for example, in remote cutting of wellheads in the case of blow-outs and well fires. It has also been used by the marine salvage industry. Recent development work has been carried out by installation contractors to produce tools and trials of sub-sea equipment using diamond wire to cut larger diameter carbon steel tubulars (i.e. larger than well casing).

The positive advantages of this method is the confidence it produces by having a fully successful through section cut proven prior to lift operations. It has been demonstrated that even under a large axial compressive load, a typical jacket main tubular member can be successfully cut. One of the reasons for this may be the progressive reduction in diamond-bead diameter as the cut proceeds across the section which is greater than the kerf closing rate due to stress relief in the tube.

Perhaps the greatest potential of diamond wire cutting lies in its development to cut the relatively larger wall section thicknesses found in CGS platforms. These are typically up to 1.5m of heavily reinforced concrete. In quarrying however, including igneous rock, thicknesses of 2 to 3 metres over 5 to 10 meter kerf length are routinely cut. The greatest problem in surface operation in air is cooling of the kerf faces. This would not be a problem underwater where productivities might actually be higher.

Shaped Charge

Shaped Charge (or more correctly Lined Cavity Explosive Linear Charge) devices have been extensively and successfully used in offshore projects [5; 9]. They have been used mostly to cut well-casing below mudline but also in many cases to cut tubulars in smaller jackets.

For major jacket structures, particularly of the third generation design which have wall thicknesses of 100mm or more in the lower main tubulars, the quantity of explosive and therefore the size and complexity of the cutting devices may be considerable. The explosive energy releases and blast hydraulic overpressures represent major problems and dictate very large stand-off distance for supporting vessels and marine plant such as SSCVs. This means that blasting with crane support is not possible and dictates a temporary support solution.

In addition there must be care taken to examine the environmental impact of multiple major underwater blasts on the pellagic and benthic fauna and biota. Economically valuable fish stocks must not be damaged and unfortunately the swim bladder of a fish is very sensitive [11].

Safety issues are of very great concern from the point of view of electromagnetically induced accidental detonation and because of the great hazard involved in diver intervention to deal with misfires.

Shaped Charges, however, will have considerable value as one of the toolbox of cutting techniques which will be applied in various combinations and sequences. Their greatest use maybe in precutting small to medium tubulars of jackets in preparation for a final phase of cutting based on one of the other techniques. It does not seem likely that shaped charge (or other) explosive techniques can be successfully applied to heavily reinforced concrete platform legs. The jet-cutting effect of cavity charge on a heterogenous material such as concrete is very much less than on steel plate and the implied magnitudes of explosive energy are very much greater.

High Pressure Water and Grit Erosion [6]

Ultra High Pressure Water at Grit Erosion has been used successfully for the internal cutting of jacket tubulars, piles and well casings but as yet us a technique with limited underwater experience but which offers major potential for both steel jackets and concrete substructures. The pressure required is of the order of 2500 bar together with very hard dry added abrasive such as carborundum or garnet grit. The greatest problem and area for development lies in the rate of wear of the jet delivery nozzles. Tests have proven UNPWG capable of operating productivity at 360m seawater depth and it is expected that 500m is achievable with development. For small volumes of material cut it would appear feasible to mount the equipment (i.e. cutting nozzle) on a working ROV. For larger cut volumes a working module would be deployed sub-sea with remote control from the surface.

It is reported that the technique can, as presently developed, cut up to 305mm of sandwich construction consisting of five layers of steel. Cutting of cement fill and reinforced concrete is easily achievable with no change of cutting nozzle and it is notable that cold-drawn hard prestressing wire embedded within cement is more easily cut than with a mechanical tool. Finally the advantages are that cutting involves no shock or vibration (assuming restraint) and is reasonably environmentally friendly (apart from abrasive grit deposits on the sea bed).

Sub-soil

The regulatory requirements for cutting and removal of piles and well casings prior to abandonment is to achieve cut-off between 3 and 5 metres below original mudline. Where tubular stubs have been left after all else is removed, these can be removed by internal shaped charge cutting or by an emerging technique where a hybrid tool is deployed to overshoot and drill around the tubular down to the necessary depth. Inside this baring head there is a diamond wire rope cutter (in fact two such units) which make a transverse cut, after which the off-cut is retrieved.

Toppling/Controlled Collapse

So far this technique has been applied only to the ill-fated Piper Alpha substructure. Given the degree of damage to the residual structure, safety and expediency dictated toppling as the Best Practicable Option. The application of this method to other jackets presupposes a more highly engineered and orderly approach.

In order to predict accurately failure modes, critical local members and toppling direction (i.e. direction of least work) a considerable amount of highly skilled structural engineering effort and analysis will be required. The original designers computer models may be of limited use for this purpose as they will all be linear elastic models. The realm of analysis which is required is that of non-linear plastic structural mechanics and dynamics. Techniques and software are available. Several researchers have made progress already in this direction.

What is required is to identify the minimum, safe and sufficient structure which can be achieved after making the maximum number of precuts in members. Obviously this condition will be created for a short seasonal window - let us say sufficient integrity to survive the standard summer storm. The residual plastic hinges must be accurately identified together with the toppling force vectors required to indicate reliably the failure rotation. These forces of course must be within the achievable range of the winches on marine plant which is from 50 to 150 tonnes. It is probable that shaped charge explosive cutting devices would be used on the off-rotation side together with delayed charges just below the plastic hinge zones on the rotation side.

The principle of toppling needs very careful review when contemplated for the legs or shafts of CGS platforms. The problems and methods involved in cutting through the reinforced concrete walls discussed above have to be carefully considered. The problem of major elements of debris impacting, cracking or penetrating the roof slabs and domes of the substructure storage cells is considerable from the point of view of both residual hydrocarbon release and damaged structural integrity of the base caisson. Directional control of the direction of rotation will be much more difficult to achieve (NB it is assumed that

CGS leg toppling could only be considered on the assumption of prior removal of all topsides and deck structures).

Heavy Lifting Techniques and Plant Available Fleet

Heavy lifting vessels are classified as Shear Leg Barges (SLB), Derrick Barges (DB), monohull crane vessels (MHCV), Semi-submersible Crane Vessels (SSCV) and Very Large Semi-submersible Crane Vessels (VLSSCV) (Fig. 5).

Derrick barges range from 200 to 800 tonne capacity and have either deck mounted ringer cranes or tub-mounted revolver cranes. Shear Leg Barges, which were developed originally for the marine salvage and coastal civil engineering markets (mainly by the Dutch) have fixed (i.e. non-rotating) A-frame masts supported by back stays to the stern of the hull. Capacities are typically from 500 to 1200 tonnes in European vessels, although these are some Japanese SLBs with a capacity of 3000 tonnes. Monohull Crane Vessels tend to be in the range from 500 to 1500 tonnes. Beyond this capacity, and even at this limit, MHCVs become sea-state barred, and Semi-submersible Crane Vessels become necessary for productive lift durations. SSCVs have capacities in the range 3000 to 4500 tonnes for single rotating lifts, and 4000 to 7000 tonnes for tandem fixed lifts.

There are now two Very Large Semi-Submersible Crane Vessels operating worldwide. The Heeremac vessel DB-102 (curiously named as a Derrick Barge by its original designer McDermott Inc.) can lift up to 6000 tonnes single rotating and 12000 tonnes tandem. Its competitor, the Micoperi M-7000, is rated marginally higher at 7000 and 14000 tonnes respectively.

The capacities and availability of these two VLSSCVs has since 1988 enabled the installation of a number of lift-installed jackets and large scale single integrated decks. The implication for decommissioning is that the same or newer vessels will be required for retro-lifting.

Lifting Appurtenances

One of the biggest problems which will have to be tackled to enable retro-lift removal of the first generation of topsides modules, module support frames and portions of partially removed jackets is the need to reinstall lifting appurtenances such as pad-eyes and pad-ears which have been removed (usually from lower modules to allow placement of upper modules). These are quality critical structural components which involve heavy plate fabrication and welding to high standards of workmanship and inspection. The quality of retro-fabrication and installation cannot be less than for original construction, which of course was done in the less difficult environment of a steel fabrication contractor's yard. It must not be thought that the standards of engineering and preparation for retro-lifting can be less onerous than those applied to installation. They will probably have to be higher.

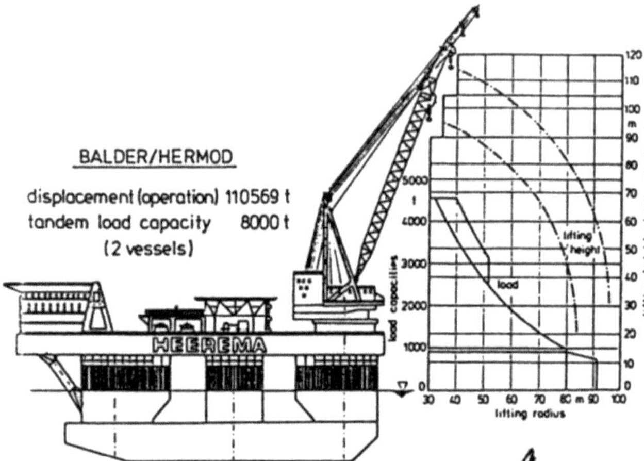
BALDER/HERMOD
displacement (operation) 110569 t
tandem load capacity 8000 t
(2 vessels)
HEEREMA
load capacities
t
lifting height
load
lifting height above water level
m
lifting radius

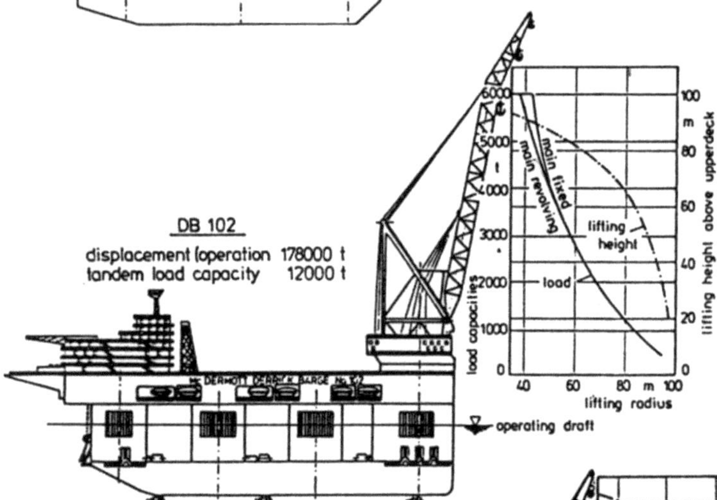
DB 102
displacement (operation 178000 t
tandem load capacity 12000 t
McDERMOTT DERRICK BARGE No. 102
load capacities
t
main fixed
main revolving
lifting height
load
lifting height above upperdeck
m
lifting radius
operating draft

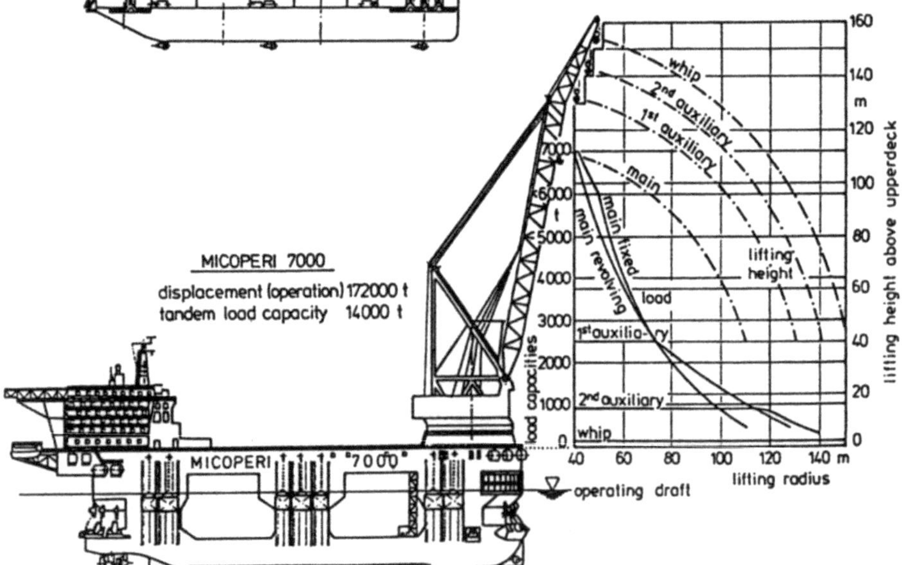
MICOPERI 7000
displacement (operation) 172000 t
tandem load capacity 14000 t
MICOPERI "7000"
whip
2nd auxiliary
1st auxiliary
main
main fixed
main revolving
lifting height
load
1st auxilia-ry
2nd auxiliary
whip
load capacities
t
lifting height above upperdeck
m
lifting radius
operating draft

Retro-Transportation

Heavy lifting operations cannot be designed without the application of careful thought to the set-down conditions and sequence.

There is no point lifting a major element from the topsides of a platform unless it can be set down on a transportation barge in a controlled fashion or set down as a single rotating lift on the deck of the SSCV. This limits the size to the single lift capacity of the strongest crane of the pair. In other words a 9000 tonne integrated deck which was originally installed as a tandem lift from a transportation barge will have to be retro-lifted onto a transportation barge. It will require to be rapidly secured and sea-fastened. Conventional fabricated and welded sea-fastening attachment will probably take too long given the likely weather window durations. Innovative forms of mechanical latching sea-fastening devices may need to be developed. For smaller lifts, say 2000 tonnes, the problem of sea-fastening will still exist because of proportionally smaller transportation barges. It would appear that it may prove more economic to carry out as much retro-transportation as possible on the deck of the SSCV.

The application of Risk Management techniques will be vital to the planning of all decommissioning operations but none more so than retro-transportation.

Retro-Load

Following successful retro-transportation of a heavy component, such as a module or section of jacket structure, to a selected shore facility, the unit must be off-loaded to the quay (or Loaded-In). Units of 2000 to 5000 tonnes are not liftable at a near shore location and so roll-off by modular trailers or skid-off on launch ways will be required exactly as operated for new-build projects. This means that the work will be done at the sites which originally operated as fabricators (provided they are still operational). It will not be viable for any new facility to be created because of the need for investment in front-end foundations, new quay or strengthened quay structures, modular trailer equipment and general civil engineering requirements. Such investment could not be contemplated unless a precontracted volume of decommissioning business could be assured.

Conclusions

The process of preparing this chapter and the course of the Symposium has produced some conclusions which the author would like to offer.

Figure 5 (opposite). Very Large Semi-Submersible Crane Vessels (VLSSCVs) and capacities (reproduced from G. Clauss et al: Offshore Structures Volume I, © Springer-Verlag 1988)

General

1. The various estimates of the cumulative cost of decommissioning lie at present in the range £5Bn to £10 Bn. This is approximately 3.3% to 6.7% of cumulative capital expenditure.

2. There are 59 structures in the UKCS which may have to be partly removed only.

3. Over the next ten years (1996–2006) only 5 or 6 major structures (probably jackets) will need to be dealt with in the Northern North Sea.

4. Generally, complete removal is not foreseen as a problem for the Southern North Sea area.

5. Professional engineers and scientists have a responsibility to be pro-active in developing commissioning plans and procedures and the communicate these in the public arena (i.e. media).

6. Public approval will not be fully measurable but disapproval will have immediate and major financial impact.

7. No new decommissioning sites would appear to be necessary or economically justifiable. The original set of construction sites will be used.

8. Even with considerable technical an public relations effort it may not be possible to persuade and gain approval from the public for ocean abyssal disposal.

9. Therefore, more expensive and more hazardous (to people and environment) solutions may have to be adopted.

10. All structures do not have the same range of options. The case-by-case approach is fully justified.

11. It is necessary to recognise that for some structures the BPEngO may be to do nothing short term and only topside removal long term.

12. The BPEngO will be driven by creative management operating on the international supply chain, incentivised by innovative fiscal measures by Government.

13. Options for decommissioning/removal must be reached on the basis of a coordinated and mutually agreed approach for all NWECS sectors (particularly UK and Norway).

14. The Deep Dumping (Abyssal Emplacement) Option is too valuable for disposal of certain innocuous major elements to be neglected. We must

not be afraid to persuade international public opinion of this and we must explain why.

Technical

15. Fitting of new lifting appurtenances to modules and jacket sections for retro-lifting will require careful engineering and high quality installation.

16. Decommissioning of moored and/or tethered structures will present no significant problems other than dock accommodation for the largest FPSO hulls.

17. Pipelines, where remaining trenched or covered do not require removal after purging and flooding. Some local exposed sections may require to be removed.

18. Modification for reuse would appear to be very unlikely by any profit-driven activity.

19. The issue of possible plant and materials recover and the economics of such activity is potentially important and needs examination.

20. All necessary cutting techniques are available but some need development for sub-sea application on a large scale.

21. Toppling or controlled collapse is a valid method of achieving the clearance requirements, but it requires very careful and rigorous analysis. It is not an option for CGS platforms.

22. Retro-lift removal by the available fleet of vessels is technically very attractive but depends on commercial issues, competition and total economics, given the small number of vessels with the highest capacities.

23. Heavy lift removal is only feasible provided the retro-transportation problems are solved i.e. set-down impact; barge size' methods of rapid sea-fastening and retro-load transfer to a quay side and site with sufficient capacity.

Recommendations

These are some recommendations which might be considered.

1. The comparisons of selected offshore platform tonnage which might be deep-sea dumped to achieve major economic advantages should be made with historic shipping loss tonnages (merchant marine and naval) since, for example, 1900.

2. The comparison including cargos and inventories should be widely publicised.

3. An attempt should be made to coordinate the Professional Engineering Institution (via the Engineering Council) to set up an Engineering Advisory Group on Decommissioning, with a pro-active brief, to emulate the recent action of the Natural Environment Research Council (see [4]) in providing independent un-biased advice and information.

4. The means of possibly increasing the available number of Hong Kong lift vessels should be studied. In particular the question of why there is no significant UK company-owned vessel should be reviewed.

5. A coordinated study of applicable and developable cutting technologies and specialist equipment should be undertaken with a view to developing the "integrated tool-box".

6. Close collaboration and coordination with the appropriate Norwegian Authorities should be pursued in order to achieve the optimum solution for the CGS structures.

Acknowledgements

The author wishes to acknowledge the assistance and encouragement provided, in particular, by Mr Graham Morrison of Health and Safety Executive, Offshore Safety Division, Aberdeen; Mr Michael Lunt, HSE (OSD), Bootle; Mr Hugh Williams of Heerema, Mr Charles Blair of Charles Blair (Aberdeen) Limited; Mr Eric Faulds of Shell UK Exploration and Production, Aberdeen, and Professor D G Gorman, University of Aberdeen, Department of Engineering. The chapter could not have been prepared without the support and resources of his then employer, the Ove Arup Partnership. The ideas and opinions expressed in this chapter are solely those of the author and do not imply approval or endorsement by any of the above mentioned individuals or organisations.

References

1. UKOOA Internet http://www.ukooa.co.uk

2. The North Sea Field Development Guide published by Oilfield Publications Ltd.

3. "Shell Shocked in Europe: The 10-day Boycott that Rocked the World" by Zachary D Lyons, Boycott Quarterly, Autumn 1995.

4. Scientific Group on Decommissioning Offshore Structures: NERC, first report, April 1996 (Internet: http://www.nerc.ac.uk)

5. Abandonment of Offshore Installations and Pipelines under the Petroleum Act 1987: Guidance Notes for Industry. Dept. of Trade and Industry Consultative Document 4 May 1995.

6. Decommissioning and Removal of Offshore Structures: Proceedings of Conference 19/20 April 1989. IBC Technical Services Ltd.

7. Removal of Concrete Platforms from the North Sea - Method and Costing Study Final Report: Ove Arup and Partners December 1977 (Confidential).

8. "Concrete Technology and Environmental Aspects Examples So Far" by Trym Edvardsson, Norwegian Pollution Control Authority. Proceedings of conference, Lillehammer, Norway 14-15 June 1994.

9. Abandonment: Technological, organisational and Environmental Challenges by Brian G Twomey and Prof. Salim Al-Hassani Paper No. H6, ONS Conference 27 - 30 August 1996, Stavanger.

10. "Death of a Platform": Artificial reef programs provide a rebirth to decommissioning structures." by Tara R Kebodeaux, Underwater Magazine Spring 1995 (http://diveweb.com).

11. A Discussion of the Key Environmental Issues Surrounding the Decommissioning and Disposal of Offshore Facilities by Sean Steere, Brown and Root Environmental published in Offshore International 1996.

12. "Report on the Environmental Impacts of Decommissioning Option" Report published by AURIS Ltd, August 1995.

13. Decommissioning—The UK Sector Dilemma" by Ron Stephen, MSc dissertation 1996 University of Aberdeen, Department of Engineering.

Appendix

Date Installed	Platform	Jacket Weight in Air (Tonnes)	1) MSF or ID Wt 2) Topsides Wt 3) Heaviest Module (Tonnes)	Water Depth (Metres)
1974	Forties A	12310	-	106
1974	Forties C	14152	-	127
1974	Auk	3414	1) 730 2) 8093 3) (1000)	82
1975	Montrose A	6500	1) 1300 2) 14000 3) (575)	88
1975	Forties B	S = 32400	-	123
1975	Forties D	S = 31000	-	122
1976	Thistle	S = 31396 (Floatout)	2) 18000	161
1976	Claymore A	12200	2) 13500 3) (2480)	110
1976	Brent A	14225	1) 1507 2) 14762 3) (807)	140
1977	Ninian South	18200	2) 26000 3) (~ 1500)	141
1977	Heather A	17000	2) 22000 3) (1200)	143
1978	Ninian North	13000	2) 15000 3) (2100)	140
1979	Tartan A	14500	1) 1400 2) 11000 3) (1300)	140
1979	Beatrice	2300	1) 2400 2) 2800 3) (1350)	46
1979	Murchison	20300	1) 5000 2) 24000 3) (2200)	156
1980	Fulmar A	12400	1) 3220 2) 24000 3) (2170)	83.5
1981	Cormorant N	20052	1) 1700 (+600) 2) 17400 3) (1700)	161
1982	Brae A	18600	2) 38000 3 (2400)	112

1982	Magnus	34400	2) 35000 3) (2420)	186
1983	Beryl B	12150	2) 24000 3) (2200)	120
1985	Alwyn NAA	18500	2) 20000 3) (4580)	130
1986	Alwyn NAB	14500	1) 3251 2) 17600 3) (2625)	130
1987	Forties SE	8000	2) 5700 3) (2886)	95
1987	Brae B	18500	2) 42000 3) (4500)	99
1988	Eider	18650	1) 4981 2) 11200 3) (2600)	157
1988	Tern	21250	2) 24000 3) (3000)	167
1990	Kittiwake	6000	1) 7500 2) 9000	85
1992	Gannet A	7750	1) 9500 2) 15600	80
1993	Nelson	8500	1) 9500 2) 22000	84
1996	Andrew	7500	2) 11000	116
1996	Captain	5000	2) 7000	104

Table A1. Table of UKCS major jacket platforms (Note: this table is illustrative, not comprehensive)

Route Operator	Fluids Carried	Length (km) Diameter (in)	Capacity	Installed
Bacton-Zeebrugge Interconnector Study Group	Gas	233 40in	-	1997
Britannia-St Fergus	Gas	195 26	21mn m³/d	1997
Cats: Everest-Teesside Amoco	Gas	397km 36in	40mn m³/d	1991/92
CMS: Murdoch-Theddlethorpe Conoco	Gas	180km 26in	21.2mn m³/d	1992/93
Cormorant South-Sullom Voe Shell	Oil	161km 36in	1mn b/d	1976, 1978
Douglas-Point of Ayr Hamilton	Gas	33.5km 20in	-	1994
Eire-UK Interconnector Bord Gais Eireann	Gas	190km 24in	18.4mn m³/d	1993
Flags: Brent-St Fergus Shell	Gas	448km 36in	31mn m³/d	1976, 1987
Forties Replacement BP	Oil	169km 36in	1,150,000 b/d	1990
Frigg-St Fergus Total	Gas	2x360km 2x32in	90mn m³/d	1974-77
Fulmar-St Fergus Shell	Gas	290km 20in	14mn m³/d	1984
Loggs: V Fields-Theddlethorpe Conoco	Gas and condensate	120km 36in	56.6mn m³/d	1987
Miller-St Fergus BP	Gas	242km 30	31.4mn m³/d	1990
Ninian-Sullom Voe Chevron	Oil	161km 36in	875,000 b/d	1976
Sage: Beryl-St Fergus Mobil	Gas	325km 30in	33mn m³/d	1990
SNIP (Scotland Northern Ireland Pipeline) Premier Transco	Gas	50km 24	-	1995

Table A2. Major trunk pipelines—UKCS (with acknowledgements to Offshore Engineer May 1996)

3

Safety and Reliability Issues of Decommissioning Offshore Structures

M J Baker

Introduction

Over the next 20 years a large proportion of the 200 or so fixed offshore oil and gas installations in the UK sector of the continental shelf will need to decommissioned, together with some of the 20 or so floating production systems. The same is true of installations in the Norwegian, Danish and other sectors.

Relatively large sums of money have already been spent on decommissioning studies for selected installations that have reached the end of their operating lives, or are close to doing so. Other, generic, studies have been carried out into specific aspects of their removal or partial removal, for example in the area of cutting and toppling of Northern North Sea jackets. This has included some basic research.

In this chapter *decommissioning* will be used in the broadest sense to include: cessation of well operations, decommissioning and removal of plant and equipment, removal or partial removal of any fixed or floating structures, removal or stabilising drill cuttings, decommissioning/removal of pipelines and making good the sea bed, along with any related on-shore activities (e.g. recycling or stabilisation of waste). Clearly not all these operations will be required for every installation.

This chapter deals with the safety and reliability aspects of the decommissioning of offshore installations, but the topics to be covered should not be considered in isolation. The health and safety of people involved in the operations depend on: the nature of the physical work being carried out, the constraints imposed by existing safety and other legislation, the need for suitable environmental protection measures, the available finance for carrying out the work, and the time constraints that are imposed, etc. As has already been seen with Shell's Brent Spar, public opinion, whether informed or uninformed also has a major role to play.

If safety and reliability issues were to be looked at in isolation it is likely that inappropriate decisions will be made. Clearly, any engineering activity, including the decommissioning of offshore structures, needs to achieve an appropriate balance between safety, environmental and financial risks. A solution which is planned to have minimal environmental consequences may (or may not) be one which poses higher than average risks to the workforce. All these factors should be taken into account before final decommissioning decisions are made.

The AURIS report commissioned by UKOOA in 1995 showed that, for many installations, the likely environmental impact is largely independent of the choice between decommissioning options. This has been the subject of an earlier session in this Symposium. If this finding is true then the choice of which decommissioning route to take hinges firmly on the safety and cost factors. To guide this choice, ways must be found of quantifying the safety risks for any given decommissioning activity. This is the main subject of this Chapter.

Basic Principles and Concepts

It might be thought that the decommissioning of offshore oil and gas installations requires a very specific approach to health and safety issues because of their unique size, location and complexity. At an operational level this is true, but at a deeper level there are no reasons why the same principles that apply to other industrial activities should not be relevant. This is indeed the case and this concept underlies the content of this Chapter.

Hazards

The starting point for any safety assessment must always be the identification of the relevant physical hazards: namely, those substances or objects with the *potential* for causing harm to people, the environment, or other objects. Physical hazards can be grouped under a number of types, as shown in Table 1. Nearly all hazards can be fitted into these categories.

PHYSICAL HAZARD TYPE	EXAMPLES
Toxic materials	chlorine, methyl iso-cyanate
Carcinogens	asbestos, vinyl chloride
Flammable materials	crude oil, methane
Explosive materials	ammonium nitrate, trinitrotoluene
Radioactive materials/sources	radon, cobalt 90
Objects with high potential energy	pressure vessel, charged capacitor
Objects with high kinetic energy	moving ship, dropped object
Dangerous microbiological substances	CJD agent, HIV virus

Table 1. Physical hazard types with examples

It should be stressed that this is a narrower definition than that given in BS 4778, which defines hazard as: 'A *situation* that could occur during the lifetime of a product, system or plant that has the potential for human injury, damage to property, damage to the environment or economic loss' (BSI, 1991).

The reason for making this distinction is that if all *physical* hazards are removed the potential for harm is reduced to zero. Unfortunately this is rarely a practical or feasible proposition, but the distinction becomes important when trying to control or reduce risks. Risk reduction may be achieved by controls on the physical hazards themselves (e.g. by reducing the amount of a flammable material stored), as well as by controls on the various initiating events and associated triggering conditions that can give rise to a release of the hazard potential (e.g. checks on the integrity of the containment vessel). Unless the distinction between physical hazards and initiating events is made, the control issues become obscured.

The wider BS 4778 definition of hazard, which encompasses the notion of initiating events, also includes the concept of psycho-social hazards. An example of the latter is an untrained person being allowed to work unsupervised in a safety-critical control room. These and other human factors issues are likely to be very important in decommissioning activities.

Finally, it should also be stressed that the presence of a major hazard does not necessarily imply high risks. Indeed, the converse may be true if major hazards are properly managed. In the offshore industry, the majority of accidents stem from routine occupational hazards and not from so-called major hazards. The Piper Alpha disaster was, of course, a notable exception.

Failure Consequences

As the result of the failure of an engineering or other system, for example during offshore decommissioning, there is likely to be a range of undesired consequences. Some of these may be a direct result of the hazard that induced the physical failure; others may result from additional hazards. The consequences may include death and injury of people (normally the workforce), permanent disablement and long-term health effects, physical damage to the engineering system itself, loss of use of the facility, loss of professional reputation, etc.

Risk

Risk is an abstract notion relating to the occurrence of undesired events in the immediate, intermediate or distant *future*. The term combines the two separate notions of the likelihood of one or more undesired events and the severity of consequences resulting from them. In the risk analysis of an engineering system it is necessary, as far as is possible:

- to identify all the undesired events that could occur

- to assess their likelihood of occurrence or frequency, and
- to estimate the nature and severity of the resulting consequences.

Unfortunately, different definitions are given for the word risk by different organisations; for example, between that given in BS 4778 ('A risk is the combination of the probability or frequency of occurrence of a defined hazard and the magnitude of the consequences of the occurrence') and that commonly used in HSE documentation ('The probability that a specified undesirable event will occur in a specified period or as a result of a specified situation'). The differences between these various interpretations of the word risk are discussed in more depth by Baker [1].

However, neither the BS 4778 nor the HSE definition of risk is sufficiently general for all purposes. A more scientific definition of risk which closely follows the sense of the BS 4778 definition will be adopted for this Chapter. The word risk is used here in the sense of 'the mathematical expectation of the consequences arising from the failure of some engineering or other system in some specified period of time (usually one year)'. This can be expressed more concisely as 'the expected consequences of failure', and can sometimes be simplified to 'the expected cost of failure' in situations where the consequences can be expressed in terms of cost. Note, however, the special (scientific/mathematical) meaning given to the word 'expected' in this context, and the fact that it is *not* the conditional expectation given the release of the hazard potential, but the unconditional expectation, that is meant.

Using this more general definition of risk, it can be seen that the units in which risks are quantified are the same as the units in which the consequences are measured.

A problem that is often raised in applying risk analysis is the apparent need of having to express risk in terms of a single quantity, generally cost (e.g. £'s), with the implication of having to put a value on life. In fact, this is *not* necessary, since risk can be considered as a *vector quantity* of the form:

$$\text{RISK} = \begin{bmatrix} \text{Expected number of dealths} \\ \text{Expected number of serious injuries} \\ \text{Expected long-term health effects} \\ \text{Expected environmental effects} \\ \text{Expected (additional) financial losses} \end{bmatrix}$$

Comparisons between two (or more) possible engineering solutions for a particular project can then be made by comparing all the components of the two (or more) risk vectors. The choice of which is the best may not be easy, since it may involve, for example, weighing a higher than average potential for serious injuries against a higher than average potential for environmental damage. It may also be necessary to weigh low initial costs against the possibility of later financial and other losses. In general, however, it should not be the engineer's or the risk analyst's sole responsibility to make the final decision, only to provide the information on which a rational decision can be made. The final choice is often a political one, to be taken by those who have been appointed, or democratically elected, to do so.

In the simplest case, when only a single mode of failure is possible, risk can be thought of as:

$$\text{RISK} = \{\text{CONSEQUENCES}\} \times \{\text{PROBABILITY OF OCCURRENCE OF THE FAILURE EVENT RESULTING IN THE CONSEQUENCES}\}$$

However, the consequences of engineering failures are themselves often highly unpredictable, and thus the calculation of risk needs to allow for the range of possible consequences. In other words, there is generally uncertainty in both the likelihood of occurrence of a failure event and in the nature and magnitude of the consequences, should the failure occur. The simple definition of risk given above is therefore rarely applicable.

Safety Legislation and the Role of the U.K. Health and Safety Executive

An earlier session in this Symposium has already addressed the general question of the legal and legislative aspects of the decommissioning of offshore installations. In this Chapter only the most important safety legislation will be covered.

As is well known, there have been dramatic changes to offshore safety legislation since the Piper Alpha disaster and Lord Cullen's report into this event. The most important has probably been *The Offshore Installations (Safety Case) Regulations 1992* which now require operators to submit a so-called safety case for various activities relating to offshore installations. One of the activities for which a safety case is demanded is for *the abandonment of fixed installations.* The use of the word abandonment here is really a misnomer, since no installation is ever likely to be *abandoned,* only the wells and remaining hydrocarbons in the reservoir to which the installation is connected.

Regulation 7 of the above, which will apply to about 200 of the 220 or so oil and gas installations currently operating in the UKCS, requires each operator not to undertake decommissioning unless:

(a) he has prepared a safety case containing the particulars specified in regulation 8 and Schedule 5;

(b) he has sent the safety case to the HSE at least 6 months before the commencement of decommissioning; and

(c) the HSE has accepted the safety case.

Regulation 8, deals with the requirements for the management of health and safety and the control of major accident hazards. All the provisions apply throughout the operating life of an installation, as well as to installation and decommissioning. Its main requirements are for:

- the provision of adequate management systems, including safety management;
- adequate arrangements for audit of such systems;
- the identification of all hazards with the potential to cause a major accident;
- risks to be evaluated, and measures to be taken to reduce the risks to persons affected by those hazards to the lowest level that is reasonable practicable.

Schedule 5 requires ten categories of information to be provided in the abandonment safety case. This includes:

- a detailed description of the facility immediately prior to the commencement of decommissioning and the operations that will be in progress;
- the numbers of persons who will be present during the various phases of decommissioning;
- arrangements for the control of well operations, including the control of well pressure, the release of hazardous substances and the avoidance of damage to sub-sea equipment by drilling equipment;
- the description of any pipeline with the potential to cause a major accident;
- particulars of plant and arrangements for the detection of any toxic or flammable gas; the detection, prevention or mitigation of fires; and the protection of persons;
- details of escape routes and means of evacuation and rescue in any emergency;
- sufficient particulars to demonstrate that the proposed arrangements, methods and procedures for:

    (a) dealing with any wells to which the installation is connected

    (b) decommissioning the installation and any connected pipelines

    (c) demolishing or dismantling the installation and any connected pipelines

  take adequate account of the design and method of construction of the installation and its plant, and reduce the risks from a major accident to the lowest level that is reasonably practicable.

This legislation, therefore, has very far reaching implications for the whole decommissioning process and the HSE is required to play a central role, both is assessing the technological feasibility and in ensuring that risks to people are reduced to a level which is as low as reasonably practicable.

In its published research strategy (1995) the HSE has set out the following research objectives in relation to decommissioning:

- To identify methods of abandonment and their associated hazards
- To identify current research on decommissioning and to identify further research requirements
- To review the engineering implications of pipeline abandonment
- To address the safety issues of large jacket toppling
- To provide technical data on subsea cutting techniques, their reliability and implications for diver safety.

From the above it can be concluded that the industry is far from knowing the answers to all the health and safety problems associated with decommissioning. Carefully planned R&D programmes are urgently needed to resolve many of these problems, and to maximise the benefits from the experience that will be gained from decommissioning some of the early installations.

Quantification of Health and Safety Risks

In trying to quantify the health and safety risks for any engineering operation two basic approaches can be used depending on the level of resource available.

The first and simplest approach is to estimate the number of man-hours of work required in each particular category of job (e.g. diving, construction, clerical, etc) and to use historical data on fatal accident and serious injury rates to estimate the 'expected' number of fatalities and injuries from this information.

The second approach, which should generally be used for the analysis of major hazard scenarios is to calculate separately the probability that each significant failure mode will occur (using either historical data or directly through the application of reliability theory) and to assess the likely consequences in terms of fatalities and serious injuries. The product of these two terms summed over all failure events then provides the 'expected' number of fatalities and injuries. This is the basis for conventional risk analyses, although this is generally based on historical failure frequencies, rather than computed failure probabilities.

Both approaches have their advantages and disadvantages and only the simpler approach will be considered here. However, the simple approach can be extended to model not only the accidents that may be 'expected' from the planned serious of events which are required to achieve full decommissioning, but also to anticipate the deviations from those plans that have a reasonable likelihood of occurring. All these issues are discussed in the following case study.

Case Study: The Brent Spar

As is well known, the Shell Brent Spar was due to have been disposed of by sinking in deep water in the North Atlantic in the summer of 1995 as this was considered to be the best practical environmental option (BPEO), and the decommission route with the lowest potential for death and injury to the workforce. However, the Brent Spar is currently moored in a Norwegian fjord awaiting a final decision, and a wider range of options is being explored.

The remainder of this section summarises the methodology that was used in a study that was carried out at the University of Aberdeen for Shell Expro to compare the human safety implications of disposal by deep sea sinking and disposal by onshore dismantling [2].

Risk analysis: Methodology

Even though the two disposal options considered here are novel in the sense that a structure of this type has not been disposed of before, none of the required operations is significantly different from normal offshore marine and construction work. Thus there is no reason to suppose that, with the necessary supervision and attention to good engineering practice, any of the disposal operations will be any more hazardous than normal offshore construction and diving work.

An estimate of the number of deaths and injuries that would result from disposal of the Brent Spar can therefore be obtained by applying industry-wide averages to the disposal options. As the detailed engineering operations are not known, this process has to be applied in a fairly crude manner to the range of engineering tasks involved. Figures 1 and 2 show these tasks for the two disposal options considered. It should also be noted that even if the detailed engineering operations were known, a more accurate analysis would be hampered by the lack of detailed statistics regarding accidents at work for the offshore industry, although significant amounts of raw data are now held by HSE and many companies.

Industry-wide averages for fatalities and injuries can be obtained from historical data. By dividing, for example, the number of fatalities that occurred in the offshore oil and gas industry over a given time period by the number of man-hours exposure to offshore risks, an estimate of the Fatal Accident Rate can be obtained. If this figure is then multiplied by the number of man-hours exposure to similar risks during disposal of the installation, an estimate of the number of fatalities that would occur during disposal is obtained.

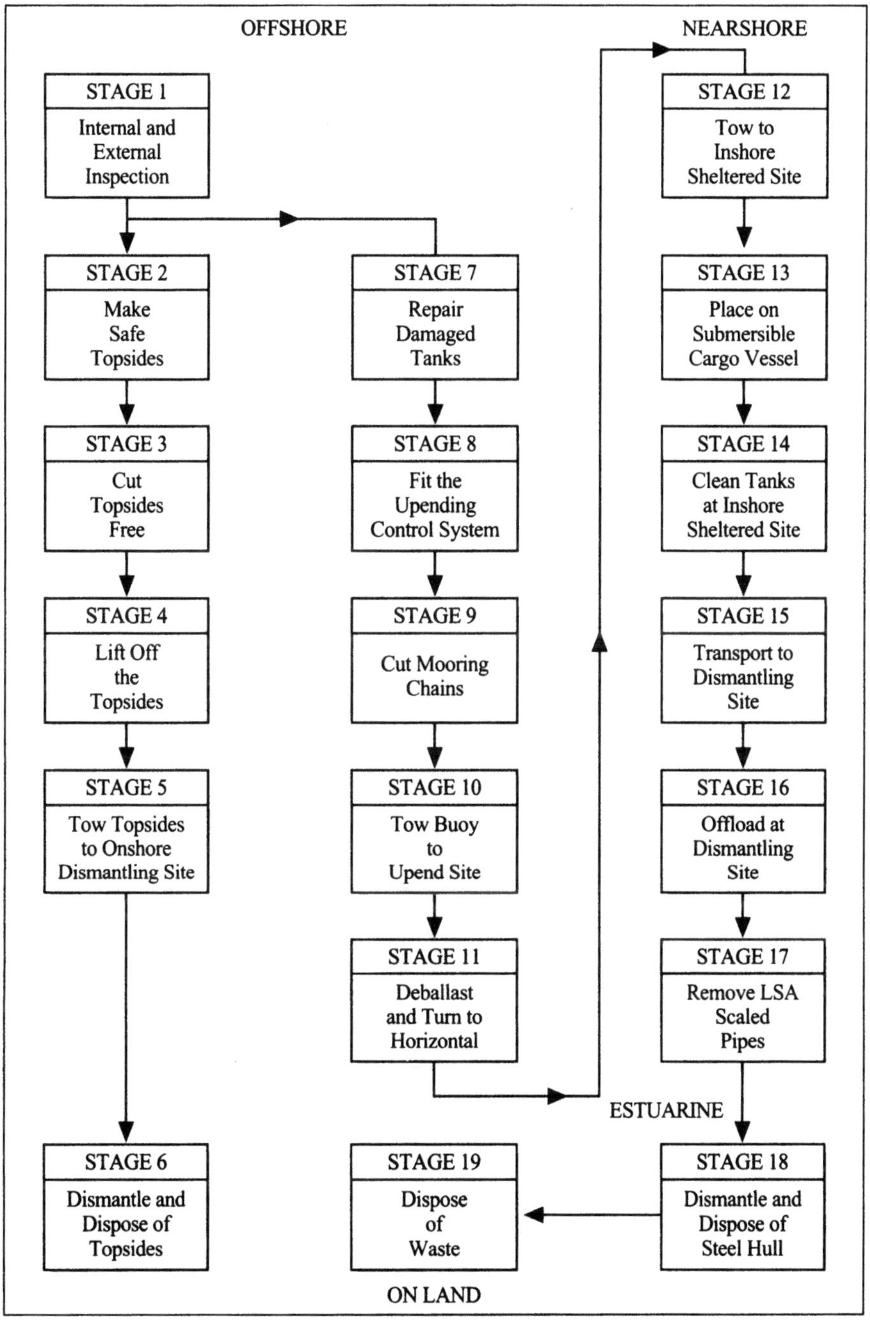

Figure 1. A summary of the main operations for the onshore dismantling option

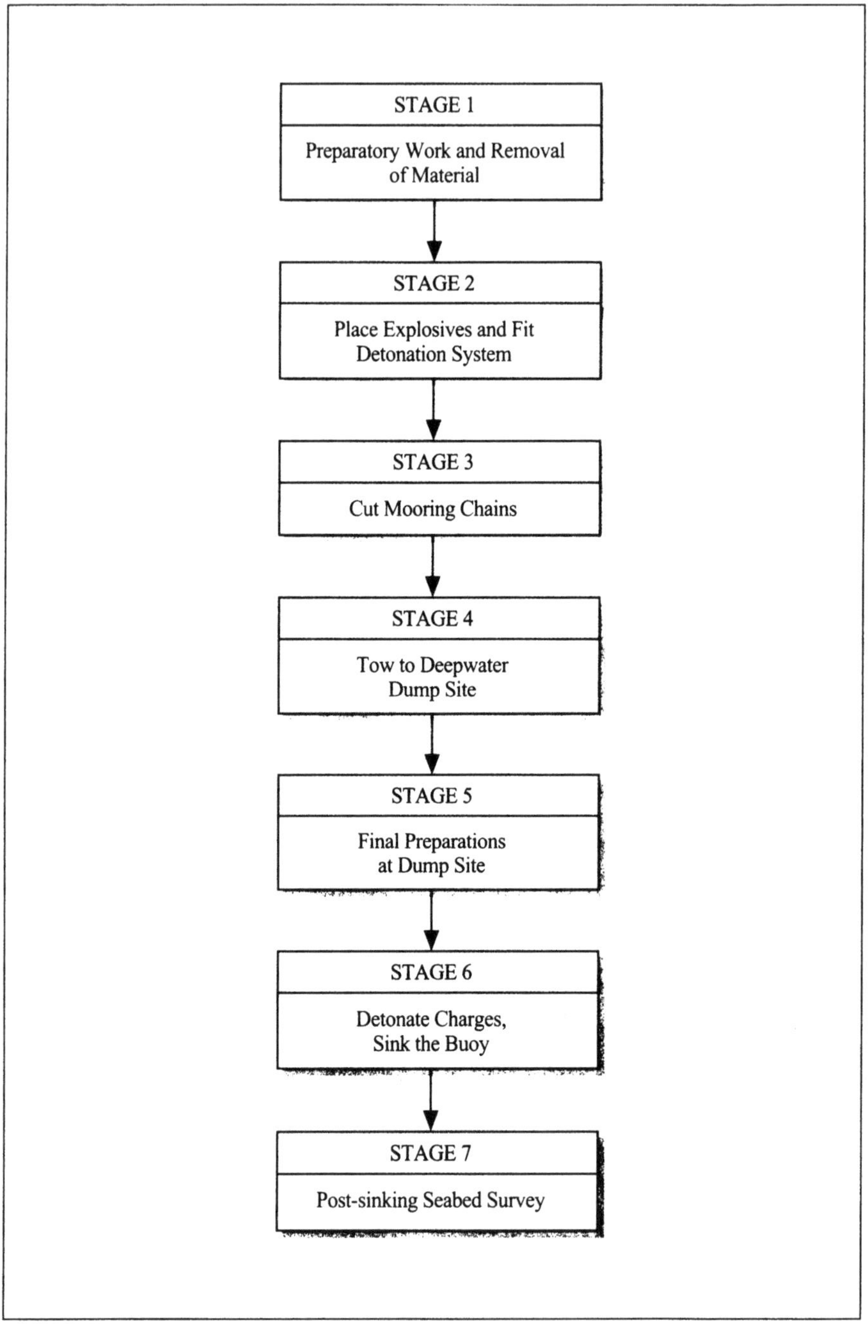

Figure 2. A summary of the main operations for the deepwater dismantling option

This technique is a simple, but useful way of estimating risks associated with engineering operations. Its main limitations are that:

1. It applies average incident rates to specific engineering tasks and does not take into account differences in risk due to task-specific hazards. Incident rates are generally obtained from an extremely wide spectrum of activities and are therefore averages taken from widely varying individual incident rates.

2. It applies historical data to future events and therefore may not take into account improvements in safety due to improvements in technology, safety management, operational supervision, etc.

3. It assumes that 'average' levels of care and supervision will be applied to the operation. The actual incident rate will depend strongly on the manner in which the tasks are supervised and carried out.

4. It does not take into account increases in risk due to the occurrence of unforeseen events. Incident rates are applied to expected exposure time, and the latter may be considerably less than actual exposure time if unforeseen events impede the execution of the engineering tasks. Unforeseen events may also alter the nature of the tasks which will need to be carried out.

5. It applies the historical rate of relatively rare events to a relatively short exposure time and is therefore likely to have a high statistical uncertainty.

Some of the above limitations may be partly offset by refining the analysis. In particular, the errors introduced by using an average obtained over a wide spectrum of activities may be reduced by splitting the tasks up into generic groupings, e.g. construction work, diving, helicopter travel. In addition, the errors arising from using historical data to future events may be reduced by taking into account long term trends in incident rates, and the errors arising from the occurrence of unforeseen events may be reduced by estimating the expected increase in risk from such events.

Incident Rates Used in Risk Analysis

The risk analyses estimated the likely number of fatalities and significant injuries resulting from each disposal option. Thus two incident rates were used, the Fatal Accident Rate and Significant Injury Rate.

Fatal Accident Rates

For the purposes of the risk analysis, the engineering tasks associated with each disposal option were divided into four categories: marine and construction work;

diving work; high risk work; near-shore and onshore work. In addition, helicopter travel was accounted for as a separate activity.

(a) Marine and Construction Work

The Department of Energy (and latterly the HSE) have been compiling accident data from offshore activities for a number of years. In addition the Norwegian authorities have been compiling similar data from activities in the Norwegian sector. Table 2 lists the number of fatalities that have arisen from construction and maintenance type activities in the Norwegian Continental Shelf (NCS) and the United Kingdom Continental Shelf (UKCS) in the past.

Location	Period	Activity	Fatalities	Man-hours ($\times 10^8$)	FAR (per 10^8 hours exposed)
NCS	1979-90	Construction	26	1.17	22
		Diving	7	0.024	291
UKCS	1980-90	Construction	65	2.43	27
		Diving	3	0.0486	62

Table 2. Summary of UK and Norwegian fatal accident data

Table 2 separates the fatalities arising from construction and maintenance type work (shown simply as 'construction' in the table) from those arising from diving work, but does not include fatalities arising from hydrocarbon related incidents. Thus the fatalities from disasters such as Piper Alpha are not included in the figures. The reason for excluding this type of fatality is that although the Brent Spar may contain some residual hydrocarbons, the risks associated with these hydrocarbons during disposal of the buoy would be extremely small when compared with the risks arising from hydrocarbons on conventional offshore installations. There would certainly not be the potential for a disaster on the scale of Piper Alpha. Thus including hydrocarbon related incidents in the Fatal Accident Rate would overestimate the risks associated with the disposal of the installation.

Table 2 also lists the Fatal Accident Rates for the UK and Norwegian sectors that were obtained by assuming an average of 1700 man-hours per man-year. Combining both sets of data to form a larger data base, an average Fatal Accident Rate of 19 per 10^8 man-hours exposed for construction and maintenance type activities is obtained.

If it is assumed that the marine and construction type activities that would need to be carried out during the disposal of the installation would be of a similar nature, and subject to exposure to similar hazards, as the construction and maintenance activities that have been performed regularly in the past, then the above figures may be used as an estimate of the Fatal Accident Rates for the disposal options.

It has been noted earlier that applying historical data to future events can lead to inaccurate estimates of risk if there have been improvements in levels of safety over the period during which the data were obtained. This is certainly the case for offshore activities which have seen a steady improvement in safety over the last ten years. Thus using an average taken over the last fourteen years is likely to overestimate the current fatality rate. Recent data suggest that the current Fatal Accident Rate for all offshore activities is in the region of 6 per 10^8 man-hours. However, as the disposal of the Brent Spar would be an operation that would involve a large amount of demolition and heavy engineering, the actual Fatal Accident Rate for disposal will probably be higher than the industry average. This is because demolition and heavy engineering are activities that have higher than average levels of risk associated with them. Hence using a Fatal Accident Rate of 6 per 10^8 man-hours in the risk analysis for the disposal of the installation is likely to produce a low estimate of the number of fatalities that would occur.

It is therefore likely that the true Fatal Accident Rate for the disposal of the Brent Spar would lie somewhere between 6 and 19 per 10^8 man-hours exposure. The risk analyses used both Fatal Accident Rates, thus giving an upper and lower estimate of risk.

(b) Diving Activities

Table 2 also lists the diving fatalities that have occurred over recent years in the North Sea. Combining the UK and Norwegian data gives an average fatality rate for the North Sea of 138 per 10^8 man-hours.

Like offshore construction and maintenance work, diving activities have seen a significant improvement in safety over the last few years. The Fatal Accident Rate for such activities is continually falling and it is estimated that the current rate is approximately 50 per 10^8 man-hours exposure.

These two rates were used in the risk analysis to give upper and lower estimates of risk associated with diving activities. As the diving personnel are also included in the marine spread the extra risk to which they would be exposed as a result of diving activities is the difference in the Fatal Accident Rate for marine and construction activities and that for diving. Thus the increase in Fatal Accident Rate is 119 per 10^8 man-hours for the upper estimate and 44 per 10^8 man-hours for the lower estimate.

(c) High Risk Activities

It was noted earlier that estimates of risk for future operations should take into account the possibility that things do not go according to plan and deviations from the planned sequence of events may be necessary. These deviations could be due to unforeseen circumstances, poorly executed or supervised operations, interference from bad weather, equipment malfunction or simply bad luck. Should such deviations occur it is possible that some high risk activities will need to be carried out. These activities are likely to have higher Fatal Accident Rates associated with them since they are unforeseen and often carried out in adverse conditions. They will not be as carefully planned as the expected sequence of tasks and may have to be carried out under conditions of stress due to time pressures, bad weather conditions, etc. The high risk activities are likely to be highly varied in nature with large variations in risk associated with them. For the purposes of the risk analysis it was assumed that, on average, these activities have Fatal Accident Rates two and a half times greater than those associated with normal offshore construction and maintenance work.

(d) Onshore Cleaning and Dismantling

From Figure 1 it can be seen that the onshore dismantling work would involve some near-shore and onshore activities. Although the precise nature of the tasks involved is not known at present, the general kind of work that would need to be carried out would be of a fairly hazardous nature. It would involve the work force entering the main body of the buoy, cleaning out the tanks and dismantling the structure. Thus the nature of the work is more akin to demolition than construction. The HSE do not record accident statistics for demolition work separately, but it is well known that this type of work is considerably more hazardous than normal onshore construction work.

In the risk analysis it was assumed that the near-shore and onshore work would have similar Fatal Accident Rates as offshore construction and maintenance type activities. Thus the upper estimate was assumed to be 19 per 10^8 man-hours exposed (approximately four times higher than normal onshore construction work) and the lower estimate was assumed to be 6 per 10^8 man-hours exposed (20 per cent higher than normal onshore construction work).

(e) Helicopter Travel

Both disposal options would involve a certain amount of helicopter travel for the work force and supervisory staff. This is a high risk activity that has resulted in many fatalities in the past. Data obtained from the years 1977-1992 indicate that there were approximately five fatalities in every 10^5 helicopter trips. In the risk analysis it was assumed that this fatality rate would apply to both the disposal options. This figure was used in both the upper and lower estimates of risk.

Significant Injury Rates

When considering risk to personnel arising from an offshore operation, it is important to consider the number of injuries that are likely to result from the operation. The HSE classify injuries as 'non-fatal major injuries' and 'over three day injuries'. In the risk analysis a significant injury was deemed to be one that resulted in three or more days off work.

(a) Marine and Construction Work

The figures from 1990/91 indicate that, for normal onshore construction work, for every fatality accident there are 213 major or over three day injuries. This figure has remained approximately constant for many years. If this ratio of significant injury to fatality is assumed for offshore marine and construction work, an upper estimate of the Significant Injury Rate of 4050 per 10^8 man-hours is obtained. The corresponding lower estimate is 1250 per 10^8 man-hours.

(b) Diving Activities

There are no equivalent figures for diving injuries involving three or more days off work. The ratio of significant injury to fatality is likely to be significantly less for diving than for marine and construction work due to the nature of diving. In the risk analysis it was assumed that Significant Injury Rates for diving are 20 times greater than the corresponding Fatal Accident Rates. This gives upper and lower estimates of 2760 and 1000 per 10^8 man-hours exposed.

(c) High Risk Activities

These activities were assumed to have significant injury to fatality ratios similar to those for normal marine and construction work. Thus the Significant Injury Rates used in the risk analysis were 2.5 times greater than those for offshore marine and construction work.

(d) Onshore Cleaning and Dismantling

These activities were assumed to have the same Significant Injury Rates as those quoted for offshore marine and construction work.

(e) Helicopter Travel

Helicopter travel has very low significant injury to fatality ratios. The likelihood of being injured in a crash but not killed is similar to the likelihood of being killed. Historical data suggests that the ratio of significant injuries to fatalities is approximately 1.5. This ratio was used in the risk analysis, giving a Significant Injury Rate of 7.5 per 10^5 helicopter trips.

Summary of Incident Rates

All the incident rates used in the risk analysis are listed in Table 3.

Activity	Fatal accident rate (per 10^8 man-hours exposed)		Significant injury rate (per 10^8 man-hours exposed)	
	Upper estimate	Lower estimate	Upper estimate	Lower estimate
Marine and construction	19	6	4050	1250
Diving	138	50	2760	1000
High risk work	48	15	10125	3125
Onshore and near onshore work	19	6	4050	1250
Helicopter travel (per 10^5 trips)	5	5	7.5	7.5

Table 3. Fatal Accident Rates and Significant Injury Rates used in risk analysis

Risk Analysis for Planned Operations

Figures 1 and 2 show the general engineering tasks that would need to be completed for the two disposal options. Although the detailed engineering work that would be required is not known at present, it is possible to estimate the approximate personnel and marine spread requirements by consideration of the general tasks that would need to be performed. By applying the average incident rates discussed in the previous section to the estimated exposure time resulting from the two disposal options, it is possible to estimate the number of incidents that are likely to occur should either option be carried out.

The estimated number of incidents for a particular activity is the product of the average incident rate for that activity and the expected exposure time to the hazards associated with that activity. For example,

Estimated number of fatalities arising from offshore marine and construction work
= Fatal Accident Rate for marine and construction work
× estimated number of man-hours spent on marine and construction work

The estimated number of incidents arising from each disposal option is then simply the sum of the estimated number of incidents from each type of activity: construction, diving, helicopter travel.

Risk Analysis for Deviations from Planned Events

Deviations from planned events were included in the risk analysis by consideration of the possible causes of such deviations and the extra work that would be required to return to the planned sequence of events. The estimated number of incidents resulting from each possible deviation is given by the probability of the deviation occurring multiplied by the estimated number of incidents that would occur if the deviation occurred. The total number of incidents occurring due to deviations from planned events is then simply the sum of the estimated number of incidents resulting from all possible deviations.

Obviously the number of possible deviations from planned events is enormous. For the purposes of risk analysis, however, it is sufficient to include only those deviations which have a reasonable probability of occurrence. For the analyses described in this document the significant deviations were chosen from consideration of the tasks involved, engineering judgement and experience of similar operations.

Examples of deviations from planned events for the two disposal options are shown in figures 3 and 4.

Summary of Risk Analysis for the Onshore Dismantling Option

When the risks associated with the planned operation are added to those associated with unexpected deviations from the planned sequence of events, an estimate of the total risk associated with the operation is obtained.

Summing the estimated number of fatalities and significant injuries for all types of activity yields the estimated fatalities and injuries for the onshore dismantling option gives:

Upper estimate of number of fatalities	=	0.088
Lower estimate of number of fatalities	=	0.030
Upper estimate of number of significant injuries	=	15.2
Lower estimate of number of significant injuries	=	4.7

These results may be interpreted as follows. The estimated number of fatalities is the average number of fatalities that would occur if the same operation were to be performed in the same manner many times. Thus if 1000 nominally identical installations were disposed of using the same onshore dismantling option it is estimated that between 30 and 88 persons would be fatally injured. Therefore, bearing in mind the uncertainties in the risk assessment process itself, it can be estimated that the probability of a fatality occurring during the onshore dismantling of the Brent Spar would be between about 3 and 9 percent.

Summary of Risk Analysis for the Deep Sea Disposal Option

As for the onshore disposal option, when the risks associated with the planned operation are added to those associated with unexpected deviations from the planned sequence of events, an estimate of the total risk associated with deep sea disposal is obtained.

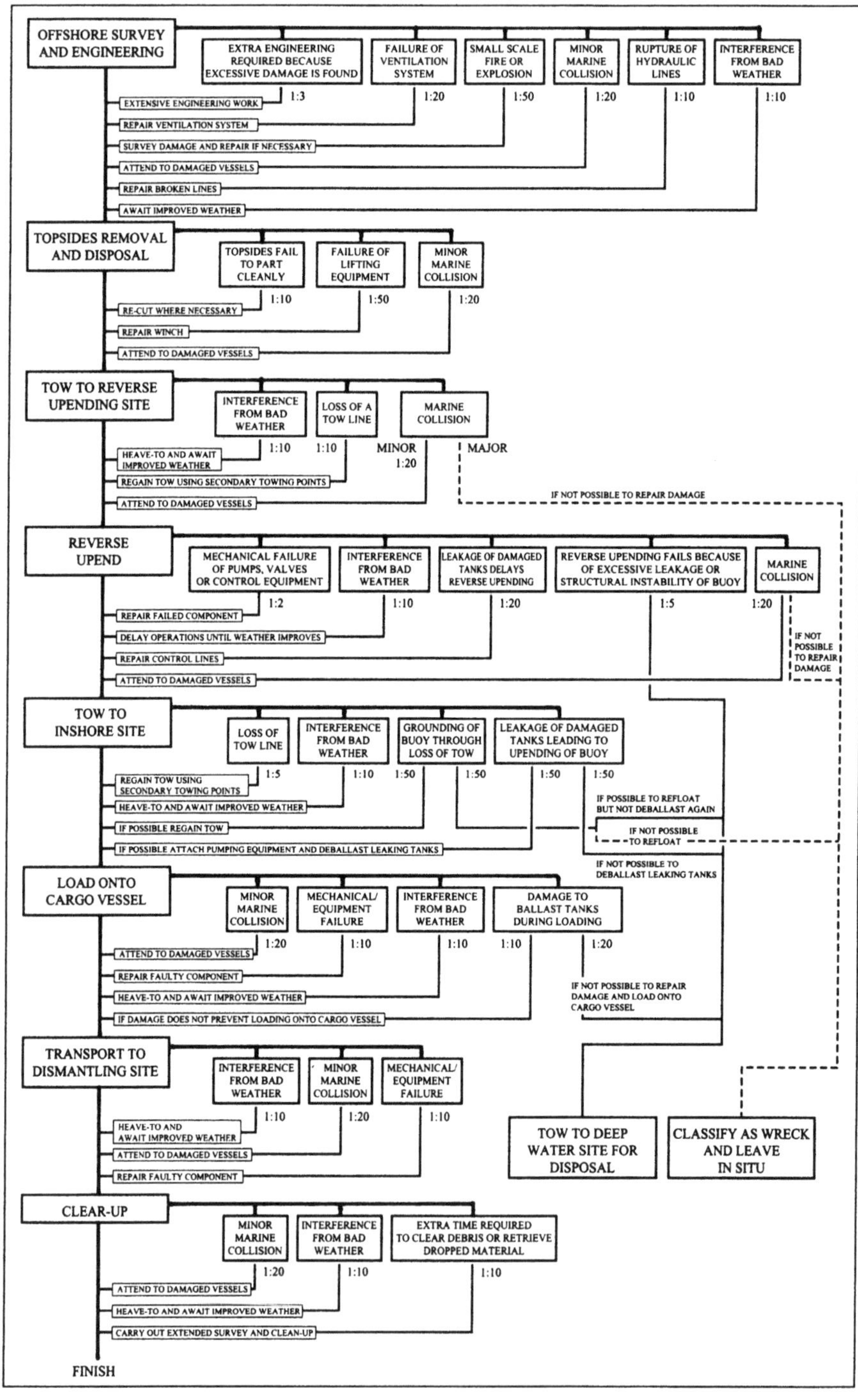

Figure 3. Flow chart of planned and unplanned events for the onshore dismantling option

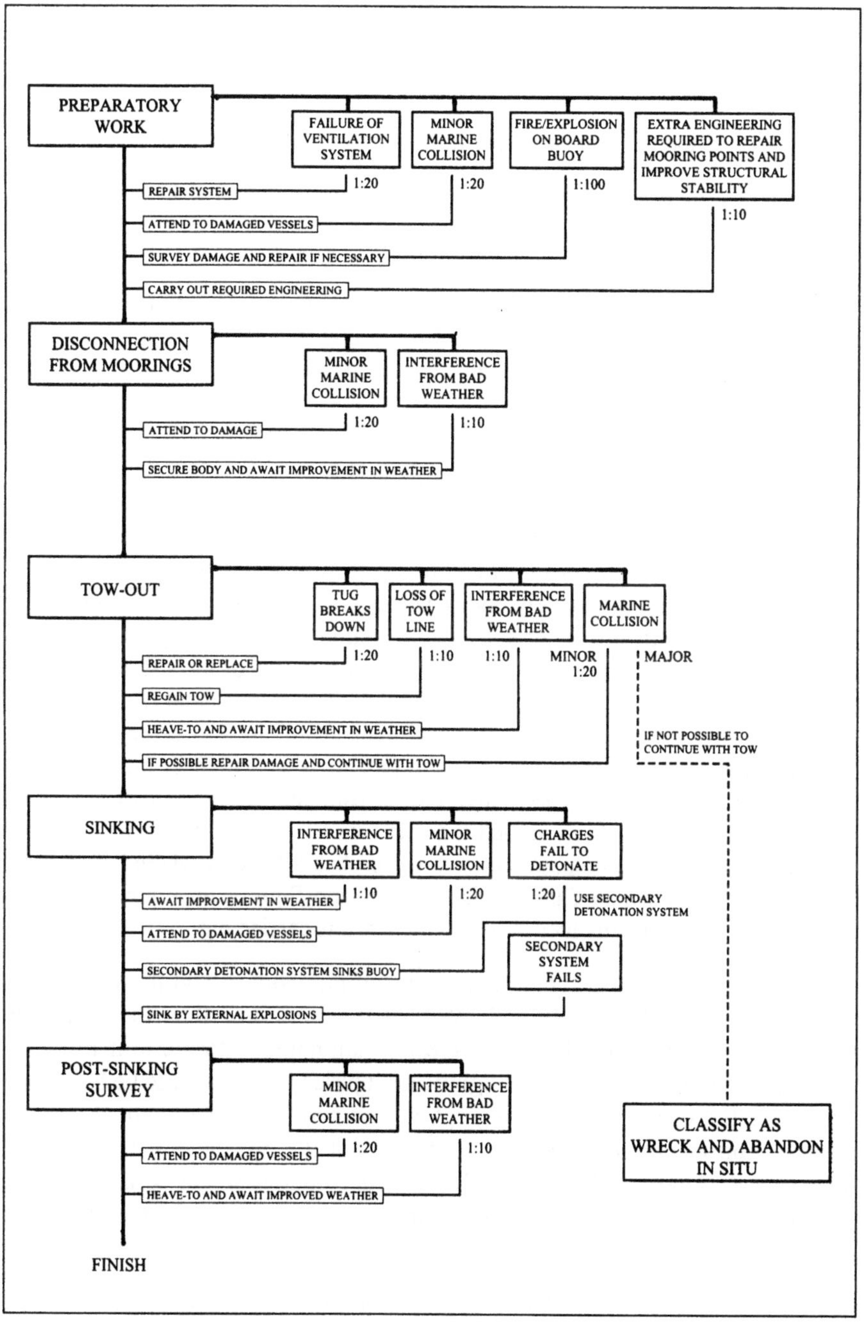

Figure 4. Flow chart of planned and unplanned events for the deepwater dismantling option

Summing the estimated number of fatalities and significant injuries for all types of activity yields the estimated fatalities and injuries for the deep sea disposal option:

Upper estimate of number of fatalities	=	0.014
Lower estimate of number of fatalities	=	0.005
Upper estimate of number of significant injuries	=	2.6
Lower estimate of number of significant injuries	=	0.7

Using the same argument as that described above, it is estimated that the probability of a fatality occurring during the deepwater disposal option would be between 0.5 percent and 1.4 percent and the number of significant injuries is likely to be between about 1 and 3.

Conclusions from the Case Study

In the case study considered, it been shown that the planned sequence of events constitutes the greater proportion of risk to the health and safety of the work force. These account for approximately eighty percent of the total estimated number of fatalities and significant injuries, with deviations from the planned events accounting for the remaining 20 percent. Thus the results of the risk analyses are strongly dependent on the assumptions made regarding Fatal Accident and Significant Injury Rates for the activities involved in the two disposal options. Two estimates of each rate were used in the risk analyses, an upper estimate based on historical data from 1979-1990 and a lower estimate based on predicted trends in incident rates. It has been argued that the true Fatal Accident and Significant Injury Rates for the two disposal options will probably lie somewhere between the two. In reality, the actual Fatal Accident Rates and Significant Injury Rates will strongly depend on the manner in which the work is carried out.

The analyses indicate that the onshore dismantling option would be significantly more hazardous to the health and safety of the workforce than the deepwater disposal option, primarily due to the greater man-hours exposure to marine and construction risks. The number of significant injuries is estimated to be between 5 and 15 for the onshore dismantling option and between 1 and 3 for deepwater disposal. Consideration of the different natures of the two options indicates that the expected number of fatalities and significant injuries would be between five and ten times higher for the onshore dismantling option than for the deepwater disposal option.

Finally, it should be noted that the conclusions reached in this case study are based on estimates of the most likely outcomes given that a particular disposal option is adopted. This is a rational basis for decision making, but is no guarantee that the predicted outcomes will be achieved. Indeed, the eventual outcomes from the choice of a particular decommissioning option may, solely by chance, be much better or worse than predicted.

References

1. Baker, M.J. (1996). Risk Modelling and Quantification. Lecture No 2, *An Engineer's Responsibility for Safety*. Hazards Forum, London.

2. Corbett, G.G., Macdonald, K.A., Baker, M.J. and Deans, W.F.(1994) Safety Assessment of the Removal and Disposal of Brent Spar. Report to Shell Expro UK.

4

Economic and Fiscal Aspects of Decommissioning Offshore Structures

A G Kemp and L Stephen

Introduction

The subject of field decommissioning has many dimensions. In this chapter the economic and fiscal aspects relating to the situation in the UKCS are examined. The specific issues analysed are (1) economic criteria for optimising the timing of decommissioning of individual fields including mothballing aspects, (2) the likely timing and aggregate expenditures on decommissioning activities in the UKCS, (3) the operation of the system of tax reliefs for the expenditures involved, and (4) decommissioning aspects of the re-use of facilities.

Economic Criteria for Optimising Timing of Field Decommissioning

In discussions of the issue of the optimal timing of field decommissioning three criteria have been considered over the last few years. These are (a) negative post-tax profits, (b) minimum margin on ongoing expenditures, and (c) maximisation of remaining net present value. These are discussed in turn.

(a) Negative Net Profits

At some stage in a field's production decline curve operating costs plus the royalty will exceed gross revenues. This might appear to be an obvious time for the investor to abandon a field: the apparently relevant costs are exceeding the gross income. Application of this criterion omits other relevant considerations. It does not take into account *all* future costs and revenues, and these should be considered. Thus, oil prices may have dipped only temporarily, causing net profits to become negative for a time, but they could subsequently become positive when oil prices rise again. Similarly, the negative net profits criterion does not take into consideration the actual decommissioning expenditures. Further, the fiscal reliefs given for these costs are not taken into account. For these reasons the negative net profits criterion alone is inadequate for determining the optimal decommissioning time.

(b) Minimum Profit Margin on Ongoing Expenditures

It is sometimes argued that production operations should produce sufficient income adequately to remunerate all the costs incurred in the activity including an acceptable profit margin. For example, the time of senior management and skilled engineers has a significant opportunity cost. It is argued that if they are to continue to be devoted to ongoing production activities, a return on the effort incurred reflecting this opportunity cost should be available. While this is certainly an arguable proposition this criterion also has shortcomings. It does not take into account *all* future costs and revenues and their timing. Operating margins could fall below desirable levels for a period, but could subsequently rise with increased oil prices. Use of this criterion on its own also ignores the decommissioning expenditures and the incidence of the fiscal reliefs relating to them.

(c) Maximisation of Remaining Net Present Value

Use of the remaining net present value criterion in principle involves consideration of *all* future costs and revenues, including the decommissioning costs themselves. It also takes into account the timing of all these costs and revenues and the relevant tax reliefs. Employment of this criterion is consistent with the maximisation of net present value when the field was initially planned for development. The discount rate that is appropriate may well not be the same as that initially used to screen the whole project. The latter may incorporate a premium for all the risks associated with a new field development. Many of these risks will have been reduced by late field life (though others may also have emerged). The appropriate discount rate should reflect the opportunity cost of the expertise and investment tied up in continued operations, and the perceived risks at the decision time in question.

Comparison of Decommissioning Dates with Negative Net Profits and Maximisation of Remaining NPVs in the UKCS

(a) Methodology

The comparative effects of the negative net profits and remaining net present value criteria on decommissioning dates in the UKCS have been examined with the aid of a computerised financial simulation model. The model incorporates a large databank containing key field data on recoverable reserves, investment, operating and decommissioning costs for the fields in the UKCS. The coverage includes (a) all the fields currently in production or under development, (b) those which will *probably* be developed in the next few years, and (c) those prospects that will *possibly* be developed in the medium/longer term.

The financial modelling operates by calculating the pre-tax returns, tax and royalty provisions and post-tax returns. The model incorporates the different royalty, Petroleum Revenue Tax (PRT) and Corporation Tax (CT) systems applied to the fields of different vintages, including the allowances for relief of decommissioning costs. The remaining net present values are calculated to the base year 1996. Three future oil price scenarios are employed, namely, $15, $18 and $23 in constant real terms. For new gas contracts the corresponding gas prices employed were 12 pence, 15 pence and 20 pence per therm in constant real terms.

(b) Results

For fields that could abandon by 2020 the results are shown in Chart 1 under the base estimate of decommissioning costs and in Charts 2 where these were 40% higher. The discount rates shown are in real terms. The year t is the year in which decommissioning would take place under the negative net profits criterion. The results are shown on the assumption that the objective of the investor is to maximise remaining real net present value at the different discount rates. Tax relief for ongoing production losses for both PRT and CT is presumed. In the case of PRT relief for losses is restricted to the profits from the field in question. Relief for decommissioning costs are in accordance with the latest legislation.

The results indicate that the optimal timing of field decommissioning under the remaining net present value criterion is frequently the same as the timing for under the net profits criterion. This is particularly noticeable at the lower levels of discount rate employed. Under the net profits criterion the decommissioning commences in the first year in which production losses start to occur. These production losses generally increase year by year as output from the field continues to decline.

At very low discount rates in the majority of cases the presence of production losses is important in determining the optimal timing date. The optimal timing depends significantly on the discount employed and on the size of the decommissioning costs. The higher are these costs the greater the benefit from their postponement. There is a wide range in field decommissioning costs in the UKCS depending on the type of producing structure involved. Thus the costs for floating platforms and fields employing sub-sea producing systems are very much less than those incurred where there are fixed platforms. This explains why there is some variation in the timing of the decommissioning for different fields at any given discount rate. Other things being equal fields with large decommissioning costs will abandon later than those with moderate costs.

The results in Charts 1 and 2 indicate that under the remaining net present value criterion in many cases the optimal decommissioning date is a year after that under the net profits criterion. The level of production losses in the first such year will be relatively small. It will then escalate as production revenue continues to fall. It follows that maximisation of remaining present value/minimisation of

Base Decommissioning Cost

	Year t	t+1	t+2	t+3	t+4	t+5	t+6	t+7	t+8	t+9<=
RNPV @ 0%	259	2	0	0	0	0	1	0	0	0
RNPV @ 2.5%	211	48	1	0	1	0	0	1	0	0
RNPV @ 5%	184	57	17	1	1	1	0	0	0	1
RNPV @ 7.5%	147	84	14	10	3	0	2	0	0	2
RNPV @ 10%	128	90	22	7	4	4	3	1	1	2
RNPV @ 12.5%	113	93	25	11	4	6	1	1	2	6
RNPV @ 15%	104	88	32	9	9	6	3	3	2	6

Chart 1: Maximisation of Remaining NPV and Optimal Decommissioning Time (Number of fields per year. CT and PRT reliefs for losses)

Decommissioning Cost increased by 20%

	Year t	t+1	t+2	t+3	t+4	t+5	t+6	t+7	t+8	t+9<=
RNPV @ 0%	258	3	0	0	0	0	1	0	0	0
RNPV @ 2.5%	205	53	2	0	1	0	0	0	1	0
RNPV @ 5%	173	62	19	4	2	0	0	1	0	1
RNPV @ 7.5%	137	82	22	9	3	4	3	0	0	2
RNPV @ 10%	114	94	25	10	5	4	1	1	2	6
RNPV @ 12.5%	101	89	31	12	10	4	2	4	3	6
RNPV @ 15%	92	80	40	16	6	8	4	3	3	10

Chart 2: Maximisation of Remaining NPV and Optimal Decommissioning Time (Number of fields per year. CT and PRT reliefs for losses)

losses in present value terms is often achieved by delaying the decommissioning for a short period. As the sizes of production losses increase the advantages of further postponement become less.

At high discount rates the advantages of postponing the decommissioning become greater because of the effect of the discounting of these costs. The results shown in Charts 1 and 2 clearly indicate that this effect becomes strong at very high discount rates. The highest rates shown in the charts are higher than those likely to be appropriate for the decommissioning decision. There is room for debate regarding the appropriate discount rate which should be employed in the decommissioning decision making. It is, of course, the case that the range of risks are not so great as those faced when the initial field development decision has to be made. On the other hand when the decommissioning decision is being made there are still substantial risks relating to (a) oil prices, (b) production rates, (c) future operating costs and (d) the decommissioning costs themselves.

The results in Charts 1 and 2 incorporate the effects of the availability of tax reliefs in the calculations for maximisation of remaining present value. It is interesting to consider what difference the availability of these reliefs makes to the pattern of the optimal timing of decommissioning. In Chart 3 the results are shown where no tax reliefs for production and decommissioning losses are available. Significant differences emerge. A considerably higher proportion of the fields now abandon at the same time as they do under the net profits criterion. The number of field decommissionings postponed beyond the year when production losses emerge is much less when no tax reliefs are available. While tax reliefs for production losses and decommissioning costs are both important it was found that

Base Decommissioning Cost

	Year t	t+1	t+2	t+3	t+4	t+5	t+6	t+7	t+8	t+9<=
RNPV @ 0%	262	0	0	0	0	0	0	0	0	0
RNPV @ 2.5%	260	1	0	0	1	0	0	0	0	0
RNPV @ 5%	256	2	1	0	2	0	0	0	0	1
RNPV @ 7.5%	251	5	0	1	1	1	1	0	0	2
RNPV @ 10%	243	8	2	2	2	1	2	0	0	2
RNPV @ 12.5%	238	11	3	1	1	1	3	1	0	3
RNPV @ 15%	228	13	8	2	2	2	2	1	0	4

Chart 3: Maximisation of Remaining NPV and Optimal Decommissioning Time (Number of fields per year. No CT and PRT reliefs for losses)

at the higher discount rates the reliefs for the decommissioning costs had the strongest influence on the optimal time. It was found that there were several cases at high discount rates where the optimal date was accelerated when tax reliefs were allowed for the decommissioning losses. This followed because the reliefs reduced the net cost to the investor to such an extent that maximisation of remaining NPV occurred at an earlier terminal date.

Detailed Analysis of Individual Field Cases

(a) Large Fields Paying Royalty, PRT and CT

In Charts 4-10 detailed results are given for 2 (real) large, mature fields in the UKCS. The base year for the calculation of remaining NPVS is 1996. In the first field it is seen from Chart 4 that the remaining NPV is always maximised by decommissioning it when profits become negative. There are always losses in present value terms from delaying the decommissioning activity. In Chart 5 the remaining NPVs from the base year 1996 are shown without tax and royalty reliefs for decommissioning and production losses. It is seen that while the remaining NPVs fall over time in both cases the availability of tax relief for the losses has a major effect in mitigating the pace of fall. The differences after a few years are very striking.

In Charts 6 and 7 the behaviour of the individual factors which determine the outcome in the case with tax relief for losses in shown. In Chart 6 the remaining present values of the key variables are shown. The remaining present value of operating costs increases over time with the extra years of expenditure. The present value of the remaining tax bill goes down over time because of the increasing impact of relief for (growing) production losses. The present value of the decommissioning costs fall because of the postponement of the timing of the expenditure. The present value of the remaining gross revenues increases over time with the extended production.

The decision to decommission depends on the interaction of these factors and how the *net* effect changes over time. This is shown in Chart 7 where it is seen that there is always a net loss from postponing decommissioning.

On the second field (Charts 8-10) the optimal time for decommissioning is seen to be the first year of negative profits when the investor's real discount rate is zero. As the discount rate is increased, however, the optimal time period is postponed. If the rate were 10% the postponement would be for 4 years. The relationships between the variables determining the outcome is now rather different. The remaining operating costs do not dominate the result so much. The benefit from postponing the decommissioning expenditure becomes relatively more important.

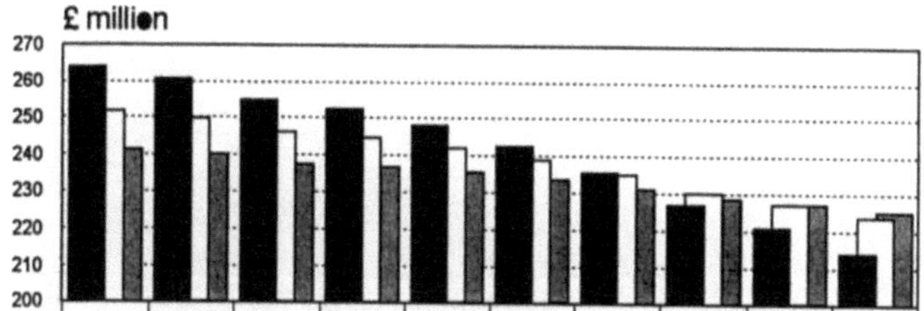

	t	t+1	t+2	t+3	t+4	t+5	t+6	t+7	t+8	>=t+9
Real NPV @ 0%	**263.7**	260.8	254.9	252.3	248.1	242.4	235.3	226.7	220.7	213.8
Real NPV @ 2.5%	**257.5**	255.1	250.4	248.4	245.2	240.9	235.6	229.3	225.1	220.3
Real NPV @ 5%	**251.6**	249.8	246	244.5	242.1	238.8	234.8	230.1	227.1	223.8
Real NPV @ 7.5%	**246.2**	244.8	241.7	240.6	238.7	236.2	233.2	229.7	227.6	225.2
Real NPV @ 10%	**241.2**	240.1	237.5	236.7	235.3	233.4	231.1	228.5	226.9	225.3
Real NPV @ 12.5%	**236.4**	235.7	233.5	232.9	231.8	230.4	228.6	226.7	225.5	224.3
Real NPV @ 15%	**232**	231.4	229.7	229.3	228.4	227.3	225.9	224.5	223.6	222.8

Years

Real NPV @ 0% Real NPV @ 5% Real NPV @ 10%

Chart 4: Large Royalty and PRT paying field
(Remaining real NPVs from 1996 base. CT and PRT relief from losses)

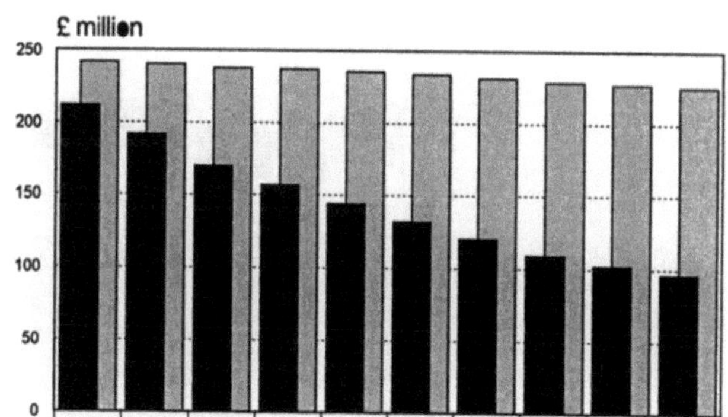

	t	t+1	t+2	t+3	t+4	t+5	t+6	t+7	t+8	>=t+9
No CT & PRT relief	**211.4**	191.1	169.2	156.1	143.4	131.3	119.7	108.7	102	95.7
CT & PRT relief	**241.2**	240.1	237.5	236.7	235.3	233.4	231.1	228.5	226.9	225.3

Years

No CT & PRT relief CT & PRT relief

Chart 5: Large Royalty and PRT paying field
(Remaining real NPVs at 10% from 1996 base. CT and PRT relief for losses)

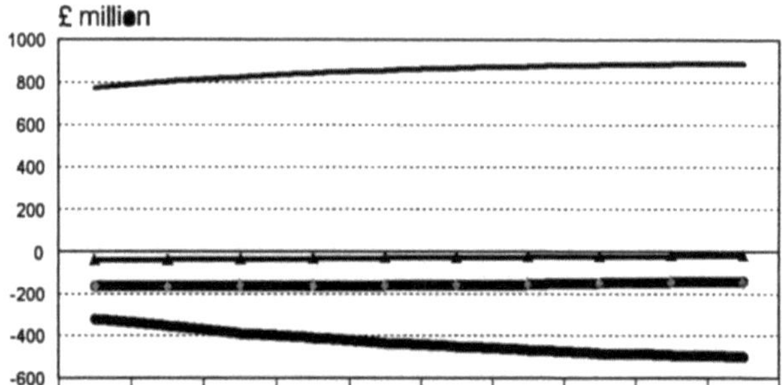

	t	t+1	t+2	t+3	t+4	t+5	t+6	t+7	t+8	>=t+9
Rem. Opex @ 10%	-321.79	-357.35	-389.95	-412.13	-432.33	-450.68	-467.36	-482.48	-492.39	-501.4
Rem. Tax @ 10%	-163.22	-165.18	-160.88	-160.55	-158.85	-156.07	-152.42	-148.09	-145.78	-143.17
Rem. Dec.cost @ 10%	-39.7	-36.09	-32.81	-29.83	-27.12	-24.65	-22.41	-20.37	-18.52	-16.84
Rem. Revenue @ 10%	769.88	802.73	825.13	843.21	857.58	868.77	877.25	883.41	887.61	890.67

Years

Rem. Opex @ 10% Rem. Tax @ 10% Rem. Dec.cost @ 10% Rem. Revenue @ 10%

Chart 6: Large Royalty and PRT Paying Field
(Remaining real NPVs at 10% from 1996 base. CT and PRT relief for losses)

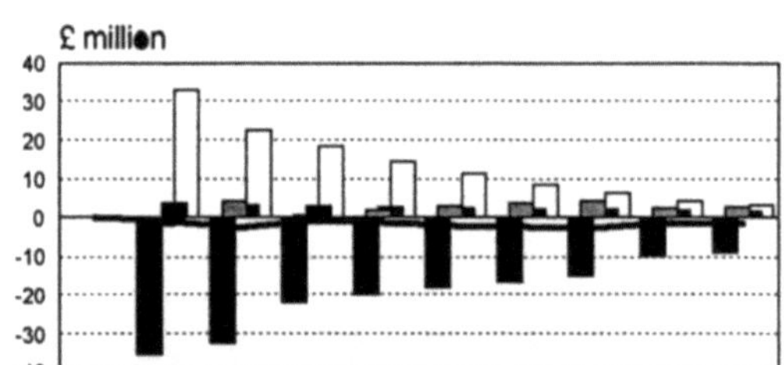

	t	t+1	t+2	t+3	t+4	t+5	t+6	t+7	t+8	>=t+9
Change in Rem. Opex	0	-35.56	-32.6	-22.18	-20.2	-18.35	-16.68	-15.12	-9.91	-9.01
Change in Rem. Tax	0	-1.96	4.3	0.33	1.7	2.78	3.65	4.33	2.31	2.61
Change in Rem. Dec.cost	0	3.61	3.28	2.98	2.71	2.47	2.24	2.04	1.85	1.68
Change in Rem. Revenue	0	32.85	22.4	18.08	14.37	11.19	8.48	6.16	4.2	3.06
Change in Net Benefit	0	-1.06	-2.62	-0.79	-1.42	-1.91	-2.31	-2.59	-1.55	-1.66

Years

Change in Rem. Opex Change in Rem. Tax Change in Rem. Dec.cost
Change in Rem. Revenue Change in Net Benefit

Chart 7: Large Royalty and PRT Paying Field
(Remaining real NPVs at 10% from 1996 base. CT and PRT relief for losses)

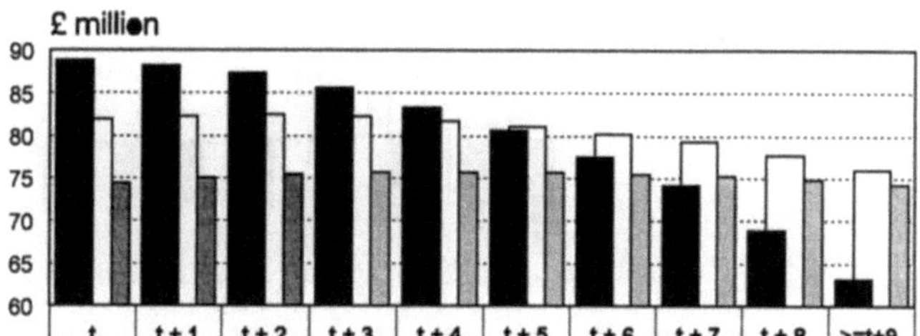

	t	t+1	t+2	t+3	t+4	t+5	t+6	t+7	t+8	>=t+9
Real NPV @ 0%	**88.8**	88.2	87.3	85.6	83.3	80.6	77.5	74.1	68.8	63.1
Real NPV @ 2.5%	85.5	**85.5**	85.3	84.6	83.5	82	80.4	78.7	75.7	72.7
Real NPV @ 5%	81.9	82.2	**82.4**	82.2	81.7	81	80.2	79.3	77.7	76
Real NPV @ 7.5%	78.1	78.6	79	**79**	78.9	78.6	78.2	77.8	76.9	76
Real NPV @ 10%	74.4	75	75.4	75.6	**75.6**	75.6	75.4	75.2	74.8	74.3
Real NPV @ 12.5%	70.9	71.5	71.9	72.1	72.2	**72.3**	72.2	72.2	71.9	71.7
Real NPV @ 15%	67.5	68.1	68.5	68.7	68.9	68.9	**69**	69	68.8	68.7

Chart 8: Large Royalty and PRT Paying Field
(Remaining real NPVs from 1996 base. CT and PRT relief for losses)

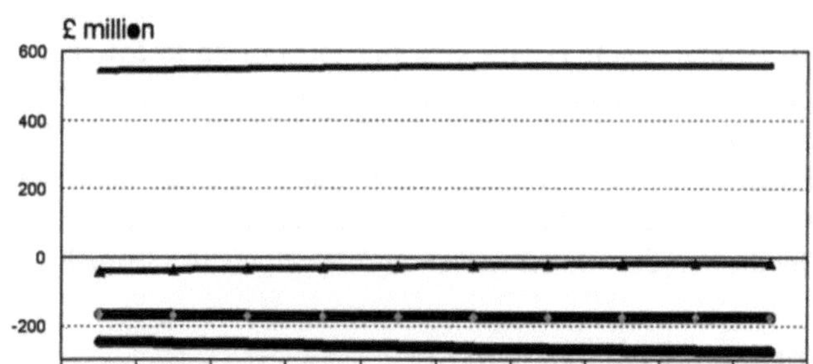

	t	t+1	t+2	t+3	t+4	t+5	t+6	t+7	t+8	>=t+9
Rem. Opex @ 10%	-245.75	-250.6	-254.49	-258.07	-261.32	-264.28	-266.97	-269.41	-272.48	-275.32
Rem. Tax @ 10%	-166.37	-169.04	-171.29	-172.93	-174.15	-175.05	-175.73	-176.25	-176.08	-175.83
Rem. Dec.cost @ 10%	-41.84	-38.04	-34.58	-31.44	-28.58	-25.98	-23.62	-21.47	-19.52	-17.75
Rem. Revenue @ 10%	543.41	547.68	550.79	553.04	554.68	555.87	556.74	557.37	557.83	558.16

Chart 9: Large Royalty and PRT Paying Field
(Remaining real NPVs at 10% from 1996 base. CT and PRT relief for losses)

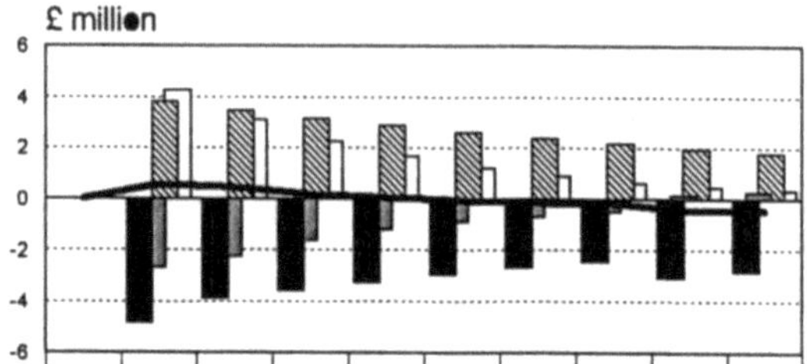

	t	t + 1	t + 2	t + 3	t + 4	t + 5	t + 6	t + 7	t + 8	>=t+9
Change in Rem. Opex	0	-4.85	-3.89	-3.58	-3.25	-2.96	-2.69	-2.44	-3.07	-2.84
Change in Rem. Tax	0	-2.67	-2.25	-1.64	-1.22	-0.9	-0.68	-0.52	0.17	0.25
Change in Rem. Dec.cost	0	3.8	3.46	3.14	2.86	2.6	2.36	2.15	1.95	1.77
Change in Rem. Revenue	0	4.27	3.11	2.25	1.64	1.19	0.87	0.63	0.46	0.33
Change in Net Benefit	0	0.55	0.43	0.17	0.03	-0.07	-0.14	-0.18	-0.49	-0.49

Years

Change in Rem. Opex — Change in Rem. Tax — Change in Rem. Dec.cost
Change in Rem. Revenue — Change in Net Benefit

Chart 10: Large Royalty and PRT Paying Field
(Remaining real NPVs at 10% from 1996 base. CT and PRT relief for losses)

(b) Small Field Paying CT Only

The next case examined was a (real), small field subject only to corporation tax (CT). From Chart 11 it is seen that the optimal decommissioning time is when profits first become negative with discount rates of up to 5%, but postponement becomes progressively more worthwhile at higher rates. At 15% discount rate remaining NPV is maximised by decommissioning in year t + 2. This field is developed with a floating production system and so the decommissioning costs are absolutely not very high. The tax reliefs are not so important on this field. The relatively modest production losses can make postponement worthwhile.

Economics of Mothballing

As fields approach the end of their economic lives, as well as minimising decommissioning costs, investors will be eager to ensure that the fixed assets are employed as productively as possible. Thus an owner of processing and transportation facilities will endeavour to find alternative uses for these assets by hiring them out to other users when excess capacity becomes available. This would normally be before the field was economically depleted.

Potential new users of these facilities may not be available at the optimal time from the viewpoint of the asset owner. A distinct possibility is that potential new

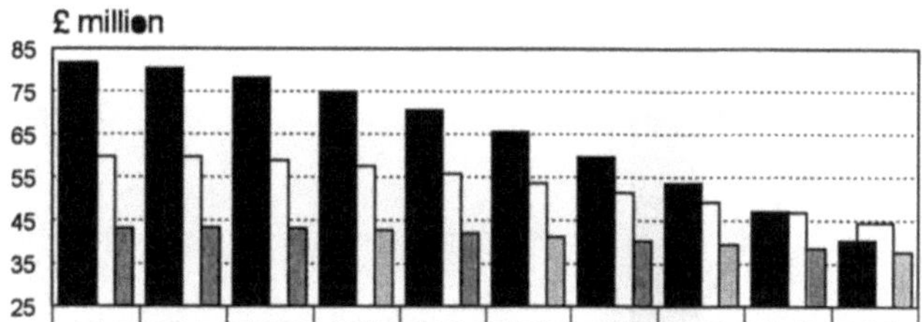

	t	t+1	t+2	t+3	t+4	t+5	t+6	t+7	t+8	>=t+9
Real NPV @ 0%	**81.6**	80.4	78.2	74.9	70.5	65.4	59.7	53.6	47	40.1
Real NPV @ 2.5%	**69.8**	69.3	68	65.9	63.1	59.9	56.4	52.6	48.7	44.7
Real NPV @ 5%	**59.6**	59.5	58.7	57.4	55.6	53.6	51.4	49.1	46.7	44.4
Real NPV @ 7.5%	50.8	**50.8**	50.4	49.6	48.5	47.2	45.8	44.4	43	41.5
Real NPV @ 10%	43.1	**43.2**	43.1	42.6	41.9	41.1	40.2	39.3	38.4	37.6
Real NPV @ 12.5%	36.4	**36.6**	36.6	36.3	35.9	35.4	34.8	34.2	33.7	33.2
Real NPV @ 15%	30.6	30.9	**30.9**	30.8	30.5	30.2	29.8	29.4	29.1	28.8

Years

■ Real NPV @ 0% □ Real NPV @ 5% ■ Real NPV @ 10%

Chart 11: Small CT Paying Field
(Remaining real NPVs from 1996 base. CT relief for losses)

users of the facilities may not be available at the time of the economic depletion of the field, but there are prospects of such business at some future date. In this circumstance the asset owner may consider whether he should maintain his facilities in anticipation of employing them.

One possibility is to cause production from his field, but then to mothball the facilities until such time as the new business materialises. The aspects of mothballing which are considered in detail here relate to the costs of such a decision. The process that is modelled covers the cessation of production decision and the consequential cost effects of the mothballing of the facilities. A schematic representation of what is modelled is shown in Chart 12. The investor is presumed to stop production when costs and royalty payments start to exceed revenues. Cessation of production costs are incurred at this time. The facilities are then mothballed. The modelling examines the direct economic consequences of the mothballing, but does not include the anticipated revenues from other activities.

Of the total decommissioning costs it was assumed that 25% related to cessation of production. Two cases were examined regarding the size of the ongoing maintenance costs required to keep the facilities in a safe condition, namely (1) 5% and (2) 10% rising to 15% of what the field operating costs had been immediately prior to cessation of production. These costs increase over time in money-of-the-day terms at the rate of inflation.

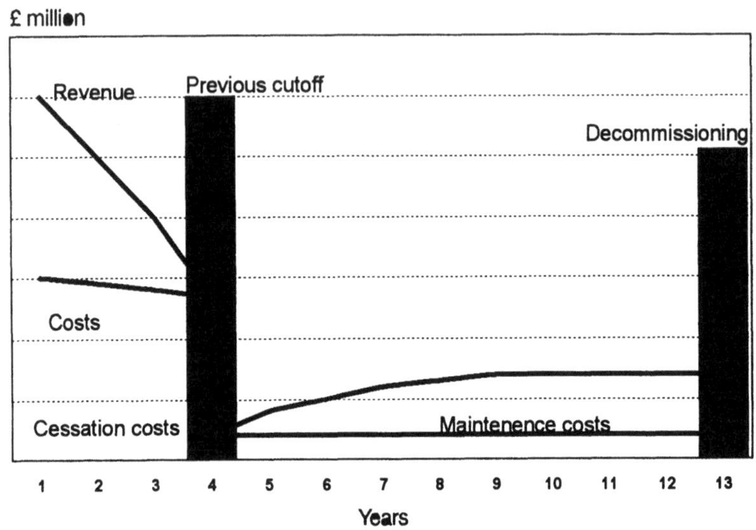

Chart 12: Mothballing model

The timing of decommissioning under the assumption that investors wish to maximise the remaining NPVs is now examined. In Charts 13-15 the pattern of the timing is shown for the bank of fields in the UK on the two assumptions regarding maintenance costs. The base year for the calculation is again 1996. The striking feature of the results is the enhanced incentive to postpone the decommissioning under the mothballing arrangements compared to the situation where production continued after profits turned negative. On a large number of fields at moderate and high discount rates a substantial postponement in the time of decommissioning becomes optimal. The annual losses are now very much less than under the scenario with ongoing production losses. This is the key explanation of the results. It is assumed in these results that tax relief is available for PRT and CT for the maintenance costs. It should be noted that PRT relief in such circumstances is available only when a firm plan for eventual decommissioning has been agreed with the UK Government.

In Chart 16 the results are shown for the case where tax relief is not available for losses during the mothballing period nor for the decommissioning costs themselves. While the results show that the incentive to postpone decommissioning is not so strong compared to the situation when tax reliefs were available, the incentive is still noticeable compared to the case of continued field production beyond the period when losses first emerge.

	Mothballed									
	Year t	t+1	t+2	t+3	t+4	t+5	t+6	t+7	t+8	t+9<=
RNPV @ 0%	250	11	1	0	0	0	0	0	0	0
RNPV @ 2.5%	182	12	3	1	4	4	4	4	5	43
RNPV @ 5%	100	9	4	2	5	4	10	7	16	105
RNPV @ 7.5%	71	6	6	2	4	3	14	10	14	132
RNPV @ 10%	64	3	6	2	4	3	15	11	16	138
RNPV @ 12.5%	61	4	4	2	4	3	14	11	19	140
RNPV @ 15%	59	4	4	2	4	3	14	11	17	144

Chart 13: Maximisation of Remaining NPV and Optimal Decommissioning Time (Number of fields per year. CT and PRT reliefs for losses. Maintenance cost = 5%)

	Mothballed									
	Year t	t+1	t+2	t+3	t+4	t+5	t+6	t+7	t+8	t+9<=
RNPV @ 0%	250	11	1	0	0	0	0	0	0	0
RNPV @ 2.5%	164	8	3	2	3	5	6	8	6	57
RNPV @ 5%	82	9	5	2	5	4	12	10	15	118
RNPV @ 7.5%	66	6	5	2	4	3	14	10	15	137
RNPV @ 10%	62	3	4	2	4	3	15	10	15	144
RNPV @ 12.5%	56	5	5	2	4	3	13	12	16	146
RNPV @ 15%	51	5	4	2	5	4	14	12	16	149

Chart 14: Maximisation of Remaining NPV and Optimal Decommissioning Time (Number of fields per year. CT and PRT reliefs for losses. Maintenance cost = 5%. Decommissioning cost increased by 20%)

Mothballed

	Year t	t+1	t+2	t+3	t+4	t+5	t+6	t+7	t+8	t+9<=
RNPV @ 0%	253	9	0	0	0	0	0	0	0	0
RNPV @ 2.5%	226	13	4	3	3	3	1	0	1	8
RNPV @ 5%	185	13	10	5	6	5	5	3	3	27
RNPV @ 7.5%	142	17	11	9	8	5	14	9	6	41
RNPV @ 10%	109	9	7	16	10	9	13	12	13	64
RNPV @ 12.5%	93	10	7	5	9	8	17	14	17	82
RNPV @ 15%	76	9	12	8	9	5	18	12	17	96

Chart 15: Maximisation of Remaining NPV and Optimal Decommissioning Time (Number of fields per year. CT and PRT reliefs for losses. Maintenance cost increasing from 10% to 15%)

Mothballed

	Year t	t+1	t+2	t+3	t+4	t+5	t+6	t+7	t+8	t+9<=
RNPV @ 0%	260	1	0	0	0	1	0	0	0	0
RNPV @ 2.5%	170	4	4	0	2	4	7	8	6	57
RNPV @ 5%	87	4	4	2	6	5	12	10	13	119
RNPV @ 7.5%	68	4	4	2	4	3	13	10	15	139
RNPV @ 10%	61	4	4	2	4	3	13	12	15	144
RNPV @ 12.5%	56	4	5	2	5	3	13	11	16	147
RNPV @ 15%	51	4	4	3	5	4	13	12	14	152

Chart 16: Maximisation of Remaining NPV and Optimal Decommissioning Time (Number of fields per year. No CT and PRT reliefs for losses. Maintenance cost = 5%. Decommissioning cost increased by 20%)

Detailed Analysis of Individual Field Cases

The optimisation of the timing of decommissioning given the mothballing option was examined in detail for a selection of fields. The calculations again refer to real fields in the UKCS. The results are discussed below.

(a) Large Fields Paying Royalty, PRT and CT

On the first field examined it is seen from Charts 17-19 that, with maintenance costs at 5% of the operating costs just prior to cessation of production, there is now some incentive to postpone the decommissioning work. There was no such incentive in the absence of the mothballing option. Although there are no revenues the ongoing costs are sufficiently low to change the balance of the factors determining the decision. At real discount rates of 10% and above the optimal time is no less than 9 years or more beyond the year of cessation of production.

The optimal timing is quite sensitive to the size of the ongoing maintenance costs. When they increase from 10% to 15% of end-of-field-life operating costs the incentive to postpone is seen from Charts 20-22 to be much less strong. For example at 10% real discount rate the optimal timing is in year t + 2 compared to beyond t + 9 under the lower level of maintenance costs.

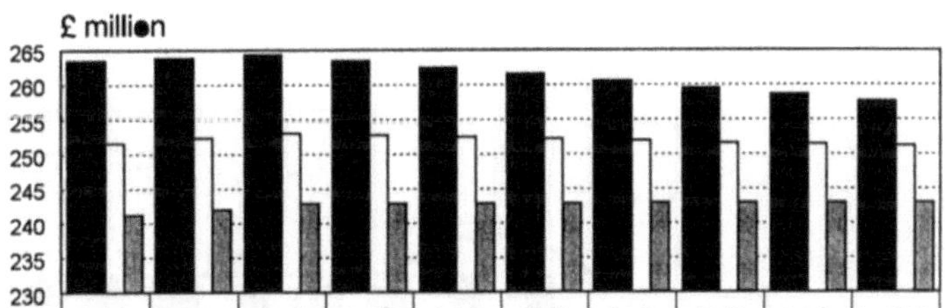

	t	t+1	t+2	t+3	t+4	t+5	t+6	t+7	t+8	>= t+9
Real NPV @ 0%	263.5	263.9	**264.4**	263.4	262.4	261.5	260.5	259.5	258.6	257.6
Real NPV @ 2.5%	257.3	257.9	**258.5**	258	257.4	256.8	256.3	255.8	255.3	254.8
Real NPV @ 5%	251.5	252.3	**253**	252.7	252.4	252.2	251.9	251.6	251.4	251.2
Real NPV @ 7.5%	246.1	247	**247.8**	247.7	247.6	247.5	247.4	247.3	247.3	247.2
Real NPV @ 10%	241.1	242	242.8	242.9	242.9	242.9	243	243	243	**243**
Real NPV @ 12.5%	236.3	237.3	238.2	238.3	238.4	238.5	238.6	238.7	238.8	**238.8**
Real NPV @ 15%	231.9	232.9	233.8	234	234.1	234.3	234.4	234.5	234.6	**234.7**

Years

Chart 17: Large Royalty and PRT Paying Field
(Remaining real NPVs from 1996 base. CT and PRT reliefs for losses. Maintenance cost 5% of opex)

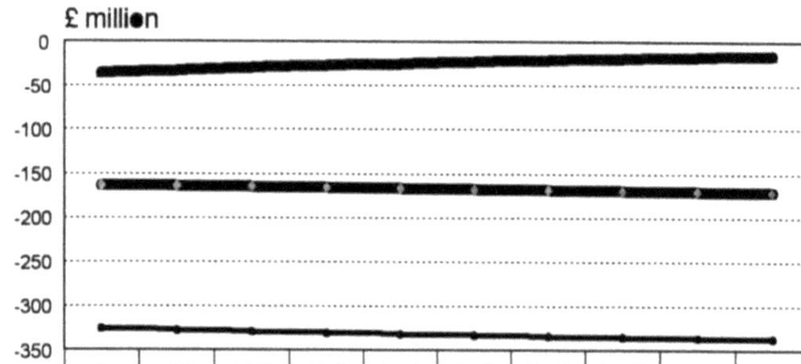

	t	t+1	t+2	t+3	t+4	t+5	t+6	t+7	t+8	>= t+9
Rem. Opex @ 10%	-325.76	-327.55	-329.17	-330.65	-331.99	-333.21	-334.32	-335.33	-336.24	-337.08
Rem. Tax @ 10%	-163.34	-163.83	-164.33	-165.51	-166.57	-167.54	-168.42	-169.23	-169.95	-170.62
Rem Dec.cost @ 10%	-35.73	-32.48	-29.53	-26.85	-24.41	-22.19	-20.17	-18.34	-16.67	-15.15

Years

Rem. Opex @ 10% Rem. Tax @ 10% Rem. Dec.cost @ 10%

Chart 18: Large Royalty and PRT Paying Field
(Remaining real NPVs at 10% from 1996 base. CT and PRT reliefs for losses. Maintenance cost 5% of opex)

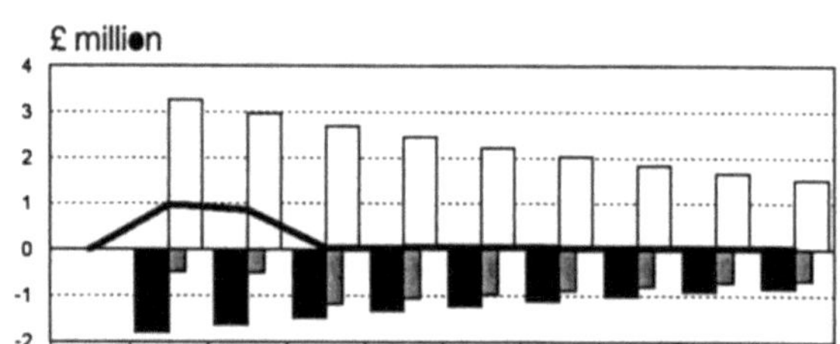

	t	t+1	t+2	t+3	t+4	t+5	t+6	t+7	t+8	>=t+9
Change in Rem. Opex	0	-1.79	-1.62	-1.48	-1.34	-1.22	-1.11	-1.01	-0.91	-0.84
Change in Rem. Tax	0	-0.49	-0.5	-1.18	-1.06	-0.97	-0.88	-0.81	-0.72	-0.67
Change in Rem. Dec.cost	0	3.25	2.95	2.68	2.44	2.22	2.02	1.83	1.67	1.52
Reduction in Net cost	0	0.97	0.83	0.02	0.04	0.03	0.03	0.01	0.04	0.01

Years

Change in Rem. Opex Change in Rem. Tax Change in Rem. Dec.cost Reduction in Net cost

Chart 19: Large Royalty and PRT Paying Field
(Remaining real NPVs at 10% from 1996 base. CT and PRT reliefs for losses. Maintenance cost 5% of opex)

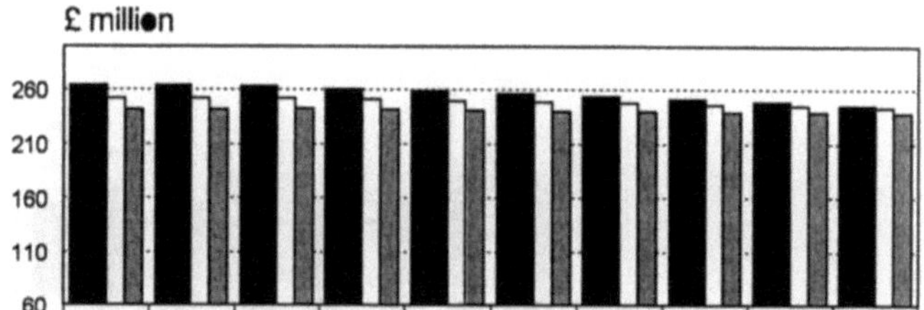

	t	t + 1	t + 2	t + 3	t + 4	t + 5	t + 6	t + 7	t + 8	>=t+9
Real NPV @ 0%	**263.5**	263	262.3	260.2	257.9	255.6	253.1	250.6	247.9	245.1
Real NPV @ 2.5%	**257.3**	257.1	256.8	255.2	253.6	251.9	250.2	248.5	246.7	244.9
Real NPV @ 5%	251.5	**251.5**	251.5	250.3	249.2	248	246.8	245.6	244.4	243.2
Real NPV @ 7.5%	246.1	246.3	**246.4**	245.6	244.8	244	243.2	242.4	241.6	240.8
Real NPV @ 10%	241.1	241.4	**241.6**	241.1	240.5	239.9	239.4	238.9	238.3	237.8
Real NPV @ 12.5%	236.3	236.8	**237.1**	236.7	236.3	236	235.6	235.2	234.9	234.6
Real NPV @ 15%	231.9	232.5	**232.8**	232.6	232.3	232.1	231.8	231.6	231.4	231.2

Years

■ Real NPV @ 0% □ Real NPV @ 5% ▦ Real NPV @ 10%

Chart 20: Large Royalty and PRT Paying Field
(Remaining real NPVs from 1996 base. CT and PRT reliefs for losses. Maintenance cost increasing from 10% to 15% of opex)

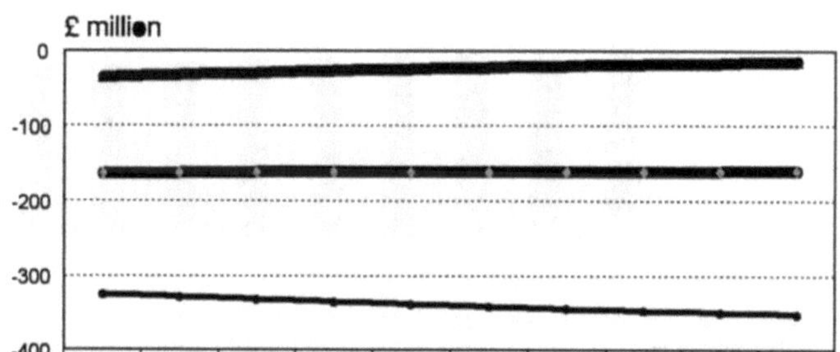

	t	t + 1	t + 2	t + 3	t + 4	t + 5	t + 6	t + 7	t + 8	>=t+9
Rem. Opex @ 10%	-325.76	-329.34	-332.76	-336.04	-339.17	-342.15	-344.99	-347.67	-350.22	-352.62
Rem. Tax @ 10%	-163.34	-162.64	-161.95	-161.92	-161.8	-161.6	-161.33	-161.02	-160.66	-160.28
Rem Dec.cost @ 10%	-35.73	-32.48	-29.53	-26.85	-24.41	-22.19	-20.17	-18.34	-16.67	-15.15

Years

Rem. Opex @ 10% Rem. Tax @ 10% Rem. Dec.cost @ 10%

Chart 21: Large Royalty and PRT Paying Field
(Remaining real NPVs at 10% from 1996 base. CT and PRT reliefs for losses. Maintenance cost increasing from 10% to 15% of opex)

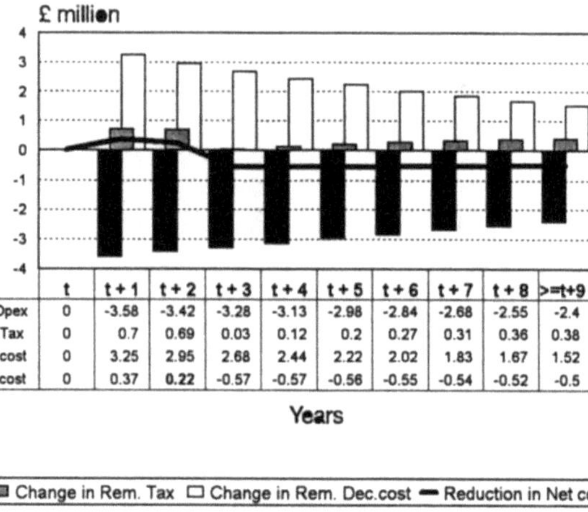

	t	t+1	t+2	t+3	t+4	t+5	t+6	t+7	t+8	>=t+9
Change in Rem. Opex	0	-3.58	-3.42	-3.28	-3.13	-2.98	-2.84	-2.68	-2.55	-2.4
Change in Rem. Tax	0	0.7	0.69	0.03	0.12	0.2	0.27	0.31	0.36	0.38
Change in Rem. Dec.cost	0	3.25	2.95	2.68	2.44	2.22	2.02	1.83	1.67	1.52
Reduction in Net cost	0	0.37	0.22	-0.57	-0.57	-0.56	-0.55	-0.54	-0.52	-0.5

Chart 22: Large Royalty and PRT Paying Field
(Remaining real NPVs at 10% from 1996 base. CT and PRT reliefs for losses. Maintenance cost increasing from 10% to 15% of opex)

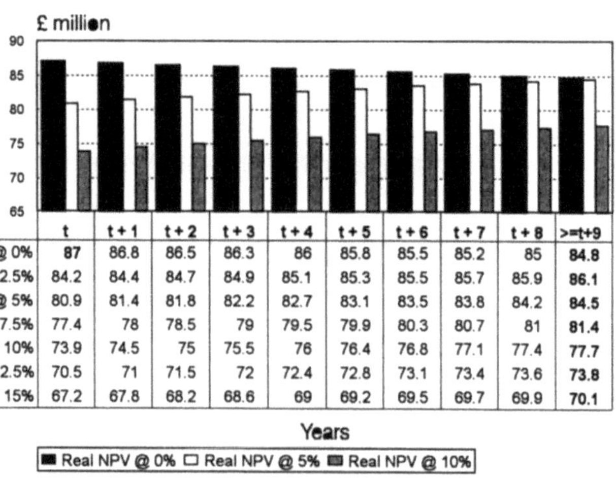

	t	t+1	t+2	t+3	t+4	t+5	t+6	t+7	t+8	>=t+9
Real NPV @ 0%	87	86.8	86.5	86.3	86	85.8	85.5	85.2	85	84.8
Real NPV @ 2.5%	84.2	84.4	84.7	84.9	85.1	85.3	85.5	85.7	85.9	86.1
Real NPV @ 5%	80.9	81.4	81.8	82.2	82.7	83.1	83.5	83.8	84.2	84.5
Real NPV @ 7.5%	77.4	78	78.5	79	79.5	79.9	80.3	80.7	81	81.4
Real NPV @ 10%	73.9	74.5	75	75.5	76	76.4	76.8	77.1	77.4	77.7
Real NPV @ 12.5%	70.5	71	71.5	72	72.4	72.8	73.1	73.4	73.6	73.8
Real NPV @ 15%	67.2	67.8	68.2	68.6	69	69.2	69.5	69.7	69.9	70.1

Chart 23: Large Royalty and PRT Paying Field
(Remaining real NPVs from 1996 base. CT and PRT reliefs for losses. Maintenance cost 5% of opex)

On the second field when maintenance costs were set at 5% of end-of-field-life operating costs the incentive to postpone decommissioning is seen (Chart 23) to be very strong. At discount rates of 2.5% and higher the optimal time is year t + 9 or more.

(b) Small Fields Paying CT Only

Two small fields subject only to CT were also examined. In both cases the maintenance costs were 5% of end-of-field-life operating costs. In the first case (Charts 24 and 25) there is no incentive to postpone decommissioning at a discount rate of 2.5% in real terms, but at rates of 5% and above the incentive is extremely strong. In the second case (Chart 26) there is seen to be *no* incentive to postpone even at very high discount rates. The key difference between the 2 cases is the very low decommissioning costs on the latter field.

Prospective Phasing of Field Decommissioning in the UKCS

The financial modelling enables estimates to be produced of the likely phasing over time of the numbers of fields being decommissioned. These are shown under two circumstances namely (a) on a field stand-alone basis, and (b) on a field interdependent basis, where the decommissioning of an installation may be postponed because it is being used to facilitate production from other fields feeding into it. The exercise was conducted to show separately the position for (a) fields in production and under development, (b) the probable and possible fields, and (c) all fields. The oil price scenarios employed were again $15, $18 and $23 in constant real terms.

In Charts 27-29 the prospective phasing of the numbers of fields which may be decommissioned is shown on a project stand alone basis. It is seen that the numbers build up substantially after 2000 to reach a peak in the period 2007-2011 when there can be 20 or more fields being decommissioned in a year.

The exercise was repeated to take account of field interdependencies. The results for all fields are shown in Chart 30. The pattern is seen to change somewhat with some postponements taking place.

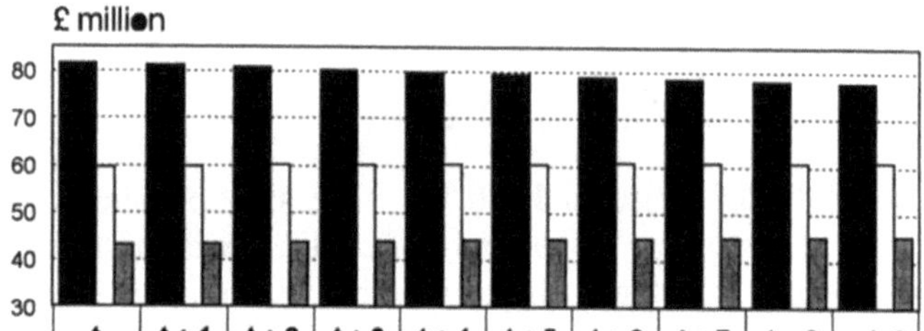

	t	t+1	t+2	t+3	t+4	t+5	t+6	t+7	t+8	>=t+9
Real NPV @ 0%	81.6	81.2	80.7	80.3	79.8	79.4	78.9	78.5	78	77.6
Real NPV @ 2.5%	69.8	69.8	69.8	69.7	69.7	69.6	69.6	69.5	69.5	69.4
Real NPV @ 5%	59.6	59.8	60	60.1	60.3	60.4	60.6	60.7	60.8	60.9
Real NPV @ 7.5%	50.8	51.1	51.3	51.6	51.8	52	52.2	52.4	52.6	52.8
Real NPV @ 10%	43.1	43.4	43.7	44	44.3	44.5	44.7	44.9	45.1	45.2
Real NPV @ 12.5%	36.4	36.8	37.1	37.4	37.6	37.8	38	38.2	38.4	38.5
Real NPV @ 15%	30.6	31	31.3	31.5	31.8	31.9	32.1	32.3	32.4	32.5

Years

■ Real NPV @ 0% □ Real NPV @ 5% ■ Real NPV @ 10%

Chart 24: Small CT Paying Field
(Remaining real NPVs from 1996 base. CT reliefs for losses. Maintenance cost 5% of opex)

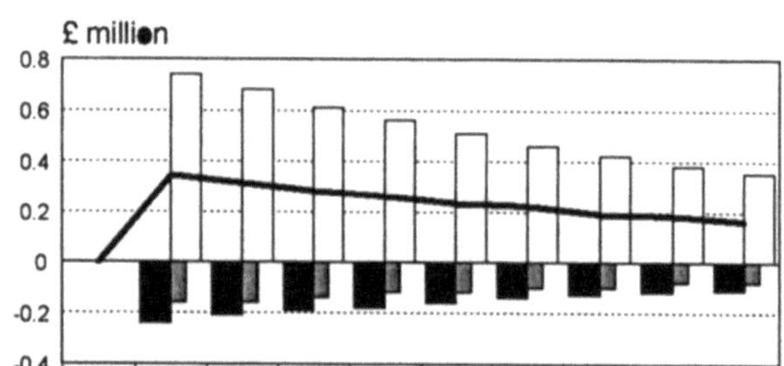

	t	t+1	t+2	t+3	t+4	t+5	t+6	t+7	t+8	>=t+9
Change in Rem. Opex	0	-0.24	-0.21	-0.19	-0.18	-0.16	-0.14	-0.13	-0.12	-0.11
Change in Rem. Tax	0	-0.16	-0.16	-0.14	-0.12	-0.12	-0.1	-0.1	-0.08	-0.08
Change in Rem Dec.cost	0	0.74	0.68	0.61	0.56	0.51	0.46	0.42	0.38	0.35
Reduction in Net cost	0	0.34	0.31	0.28	0.26	0.23	0.22	0.19	0.18	0.16

Years

■ Change in Rem. Opex ■ Change in Rem. Tax □ Change in Rem. Dec.cost — Reduction in Net cost

Chart 25: Small CT Paying Field
(Remaining real NPVs at 10% from 1996 base. CT reliefs for losses. Maintenance cost 5% of opex)

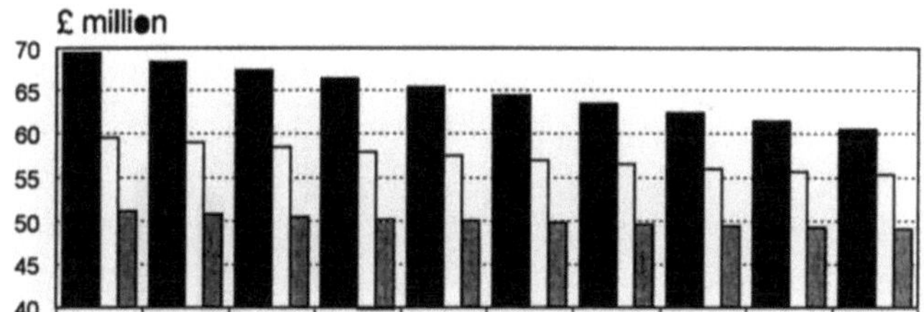

	t	t+1	t+2	t+3	t+4	t+5	t+6	t+7	t+8	>=t+9
Real NPV @ 0%	**69.4**	68.4	67.4	66.4	65.4	64.4	63.4	62.4	61.4	60.4
Real NPV @ 2.5%	**64.2**	63.5	62.7	62	61.3	60.6	60	59.3	58.7	58.1
Real NPV @ 5%	**59.5**	58.9	58.4	57.9	57.4	56.9	56.5	56	55.6	55.3
Real NPV @ 7.5%	**55.1**	54.7	54.3	53.9	53.6	53.3	53	52.7	52.5	52.2
Real NPV @ 10%	**51.1**	50.8	50.5	50.2	50	49.8	49.6	49.4	49.3	49.1
Real NPV @ 12.5%	**47.4**	47.1	46.9	46.8	46.6	46.5	46.3	46.2	46.1	46
Real NPV @ 15%	**43.9**	43.8	43.6	43.5	43.4	43.3	43.2	43.2	43.1	43

Years

■ Real NPV @ 0% □ Real NPV @ 5% ■ Real NPV @ 10%

Chart 26: Small CT Paying Field
(Remaining real NPVs from 1996 base. CT reliefs for losses. Maintenance cost 5% of opex)

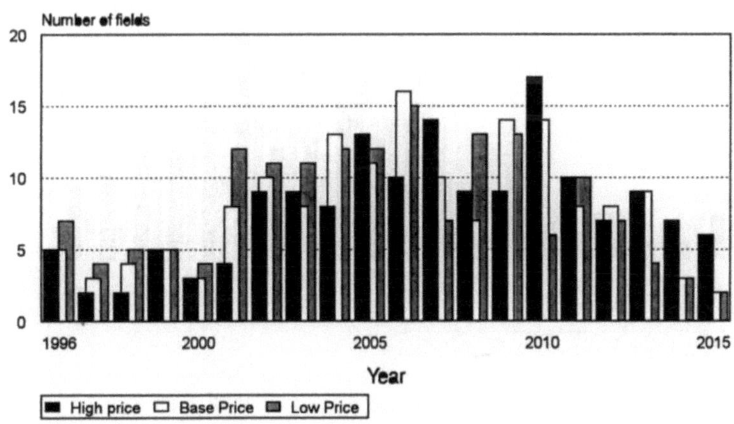

Chart 27: Number of fields decommissioned
(Fields in production and under development. Stand-alone basis)

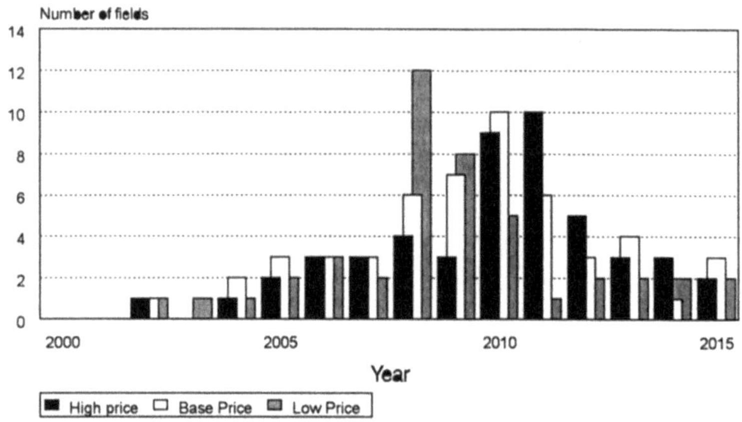

Chart 28: Number of fields decommissioned (Probable and possible fields. Stand-alone basis)

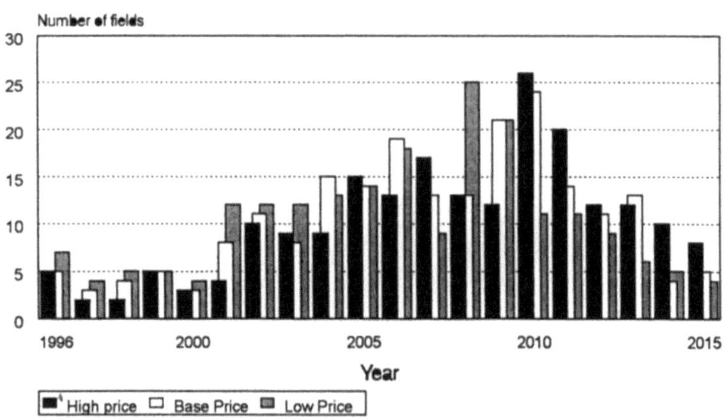

Chart 29: Number of fields decommissioned (Probable and possible fields. Stand-alone basis)

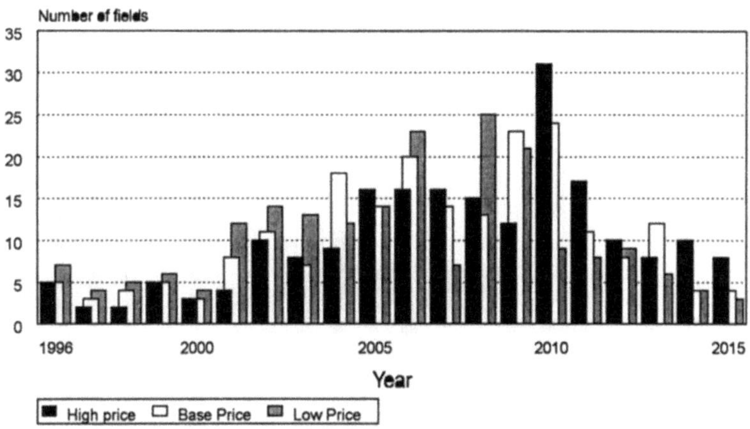

Chart 30: Number of fields decommissioned (Aggregate. Interdependent fields basis.)

Phasing of Aggregate Costs of Decommissioning

The financial modelling also permits the phasing of the costs of decommissioning to be determined. Again this exercise was conducted where the fields were considered on (a) project stand-alone basis, and (b) project interdependent basis. The prospective total and pattern of expenditures on decommissioning activities are shown in Charts 31 and 32 when the fields are viewed on a project stand-alone basis, and in Chart 33 on a field interdependent basis. On a project stand-alone basis it is seen that expenditure is relatively modest until 2001 or so, after which there is a significant increase. The level increases further from 2006 onwards, especially under the low price case. Under the high price case the levels of expenditure are postponed somewhat, and become relatively high in the period from 2010 onwards. On a project interdependent basis there is seen to be some postponement in the timing of the expenditure. The overall pattern is not dissimilar, however.

With regard to the total costs involved, for fields currently producing and under development aggregate decommissioning expenditure could be £520 million by the end of 2000, £1.9 billion by the end of 2005, £4.1 billion by the end of 2010, and £6.9 billion by the end of 2015. If future fields are also included in the calculation the figures for all fields are £4.76 billion by the end of 2010 and £7.9 billion by the end of 2015.

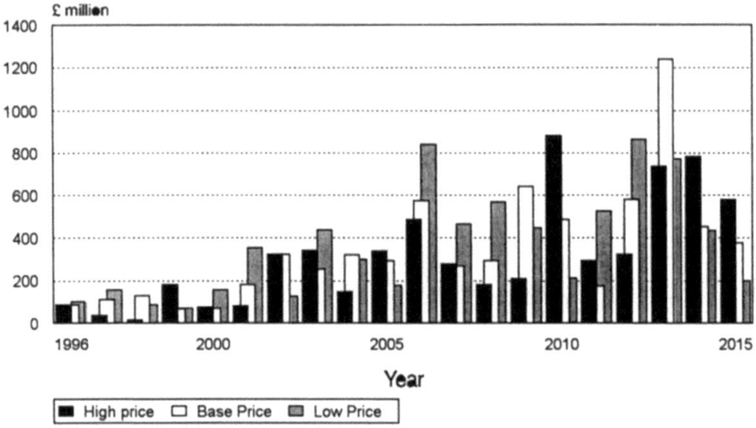

Chart 31: Real decommissioning cost (1996)
(Fields in production and under development. Stand-alone basis)

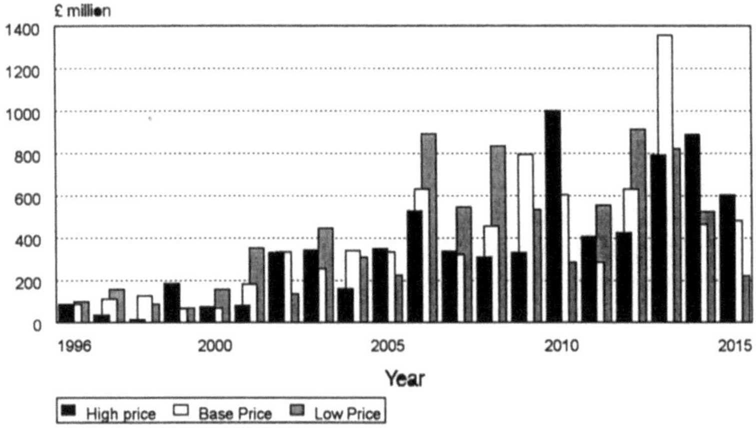

Chart 32: Real decommissioning cost (1996)
(All fields. Stand-alone basis)

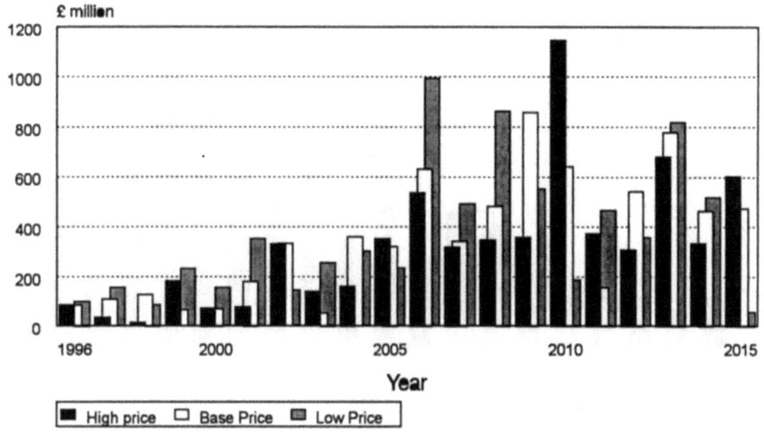

Chart 33: Real decommissioning cost (1996)
(All fields. Interdependent fields basis)

Effect of Reduction on Field Operating Costs on Timing of Decommissioning

The above analysis was undertaken employing the current best estimates of ongoing field operating costs. In recent years progress has been made in reducing these costs. In Charts 34-36 the effects of a reduction of 20% in these costs on the timing of decommissioning are shown under the 3 oil price scenarios. There is a perceptible postponement in the timing under all 3 cases. It is to be expected that some further reduction in operating costs will be achieved and thus this type of effect is likely to materialise.

Fiscal Relief For Decommissioning Expenditures and Their Effects

The evolution of the system of fiscal reliefs for decommissioning expenditures in the UK may be summarised as follows:

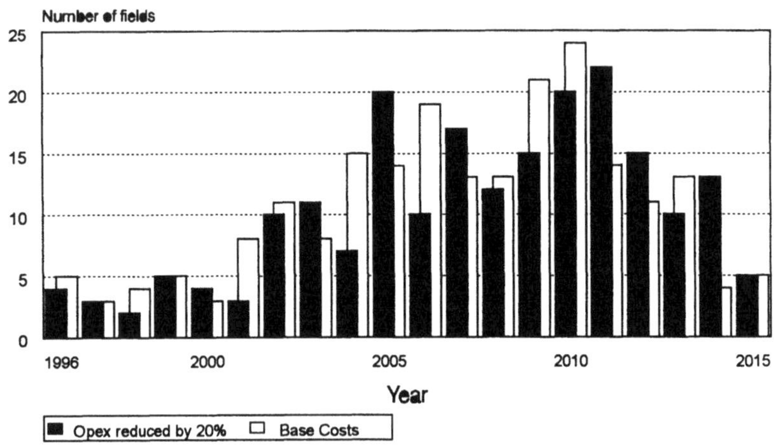

Chart 34: Number of fields decommissioned (All fields. Base price. Stand-alone basis)

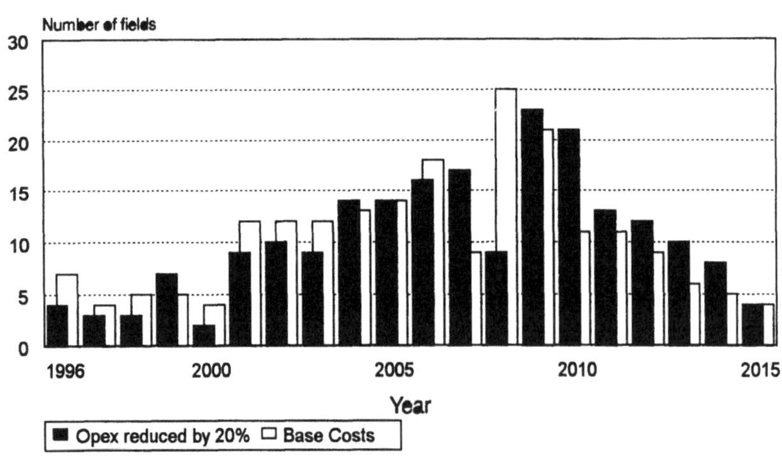

Chart 35: Number of fields decommissioned (All fields. Low price. Stand-alone basis)

(a) Pre-Finance Act, 1990

1. Royalty: on Series 2 royalties relief is available for transport and treatment costs. Royalty refunds are subject to PRT and CT. CT liability takes into account increased PRT payments.

2. PRT: decommissioning costs create losses carried back against profits (calculated before deduction for oil allowance and safeguard). Refunds receive tax free interest. CT is payable on PRT refunds for the period for which the deduction was originally given, with interest due on increased CT.

3. Corporation Tax: writing down allowances of 25% declining balance are available against other income. For ongoing investors decommissioning losses could be carried back for one year. If cessation of trade occurs losses may be carried back for up to 3 years against field income. CT refunds are given tax free with interest.

(b) Finance Act, 1990

The 1990 Finance Act introduced modifications as follows:

1. For PRT a cap was imposed on the amount of interest paid to investors in respect of PRT losses carried back to earlier periods. For losses arising from July, 1991, PRT interest plus the repayment to which it relates cannot exceed 85% of the PRT loss which is carried back to the period in question.

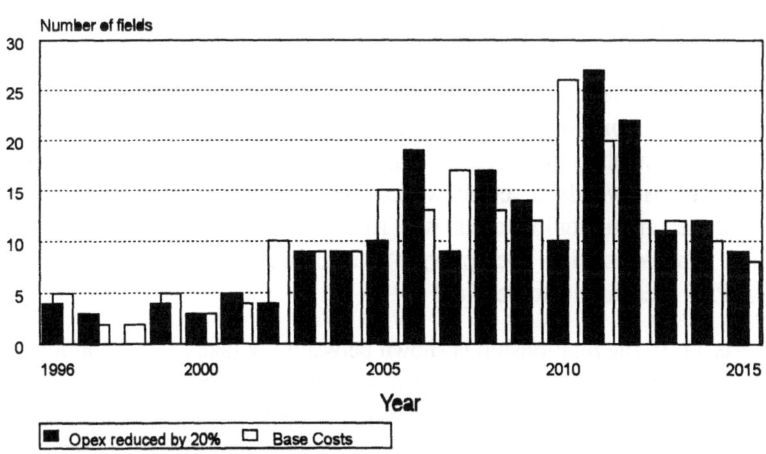

Chart 36: Number of fields decommissioned
(All fields. High price. Stand-alone basis)

2. For CT a new 100% allowance for decommissioning expenditures was introduced for expenditure incurred from July, 1991 onwards. Losses attributable to the new allowance can be carried back for up to 3 years against ring fence income. Decommissioning expenditures incurred for three years after cessation of ring fence activity will, from July, 1991, qualify for relief. From the same date PRT refunds become chargeable to CT for the period in which the PRT loss arose.

(c) Finance Act, 1991

In the 1991 Finance Act some technical changes were introduced. These give a fuller definition of decommissioning costs allowable for PRT purposes, and permit the deductibility of payments made to third parties as guarantees for approved decommissioning programmes. Defaulters' costs relating to a co-licensee are also allowed as deductions.

(d) Finance Act, 1993

For periods starting July, 1993, the cap on PRT repayments plus interest is set at 60% of the PRT decommissioning loss carried back.

The impact of the fiscal provisions for the relief of decommissioning expenditures can only be fully understood by conducting a detailed financial modelling exercise on individual fields. This has been undertaken for all the currently producing fields, those under development, and probable and possible fields in the UKCS, employing the financial model discussed above and incorporating all the fiscal provisions.

The model has been designed to calculate the tax takes and decommissioning relief on the fields. The tax take is defined as the total tax bill over the life of the field expressed as a percentage of the pre-tax field net cash flow. The decommissioning relief is defined as the tax reduction expressed as a percentage of the decommissioning costs. The results in this section concentrate on the takes and reliefs in money-of-the-day (MOD) terms. It is assumed in the results presented below that the investor has sufficient income for corporation tax purposes against which to set allowances for decommissioning losses. For PRT relief for decommissioning costs was restricted to the profits of the field in question. The legislation does provide that an unrelieved field loss can be set against PRT income in another field.

In Charts 37-39 the relationship is shown between tax takes and decommissioning relief on 262 fields. The relationship is seen to be somewhat erratic. Under the base oil price in 32% of cases the effective decommissioning relief exceeds the average tax take. There is a broad tendency for very large fields to obtain relief at a higher rate than the effective tax take, but the relationship is not precise. In some cases where the decommissioning relief is greater than the average tax take

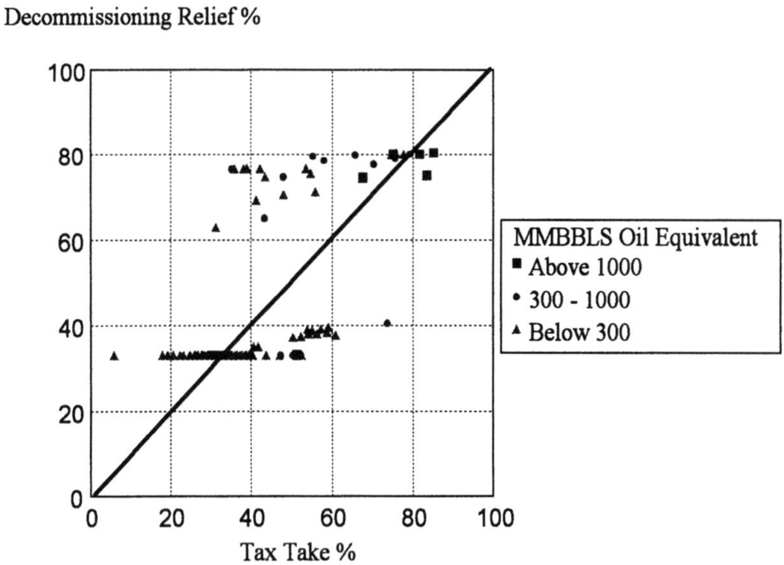

Chart 37: Relief for decommissioning (Ongoing investor base oil price)

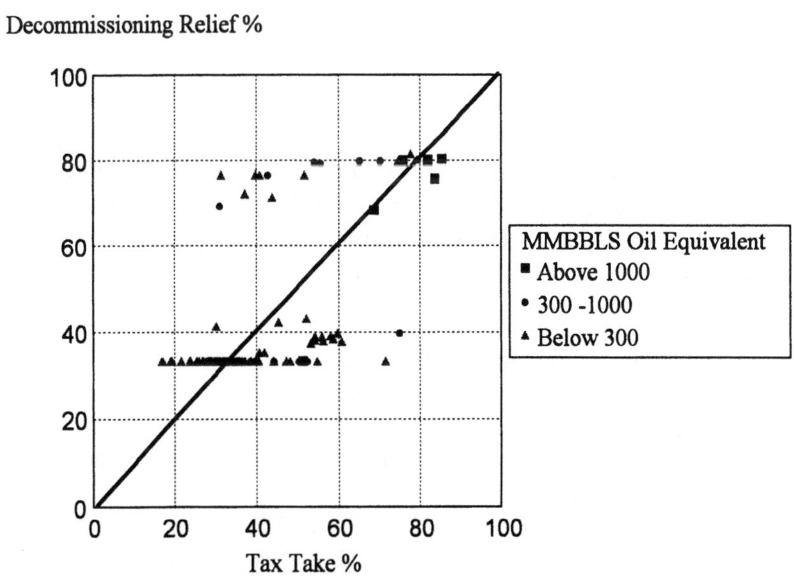

Chart 38: Relief for decommissioning (Ongoing investor low oil price)

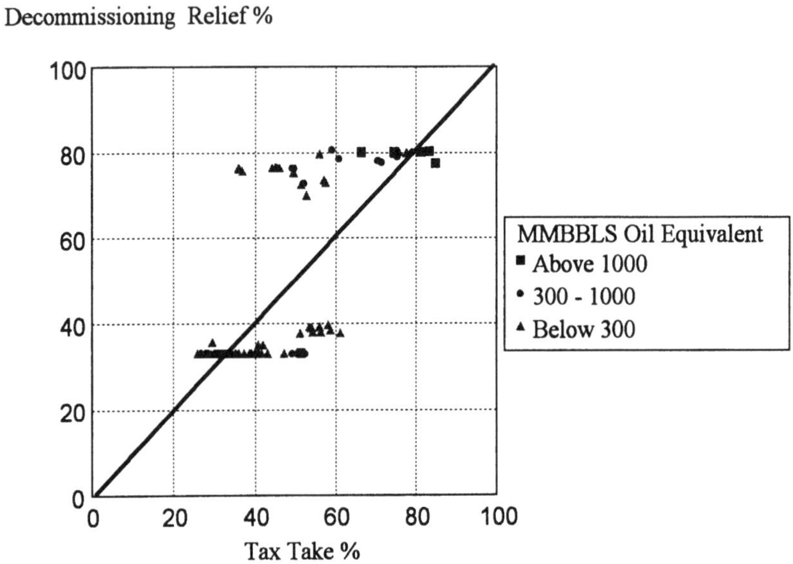

Chart 39: Relief for decommissioning (Ongoing investor high oil price)

the difference is very small, and reflects the fact that in a number of recent, generally small developments, the field has turned out to be not very profitable and full CT relief has been presumed for the decommissioning losses.

On 68% of the fields under the base oil price the effective decommissioning relief is predicted to be less than the effective tax take. It should be noted, however, that a significant number of these cases refer to the Southern North Sea where no royalty relief is available. Rather surprisingly little variation was found in the relationship between tax takes and tax relief at the different oil prices. On 215 fields the take is 33%.

There are a number of reasons for the erratic relationship displayed between decommissioning relief and tax takes. On large fields which are highly profitable on a lifetime basis full decommissioning relief will be obtainable. Thus there will be no difficulty in clawing back decommissioning losses for PRT purposes and obtaining full relief within the cap set by the legislation. With PRT at 50% the top rate of relief, including royalty, CT and interest on PRT refunds, is around 80%. In these cases the availability of interest makes a substantial difference to the overall rate of relief in MOD terms. Without it the rate of relief would be reduced by several percentage points.

In a few of the very large fields developed in the 1970's the tax take exceeds 80%, and the effective rate of relief, even at approximately 80% is below the tax take.

This reflects the fact that such fields paid significant amounts of Supplementary Petroleum Duty and a large amount of PRT at the 75% rate while relief for decommissioning expenditure was obtained at the 50% PRT rate.

On some other fields the relief for decommissioning costs exceeds the tax take because, while full relief is attained, the effective PRT take on the field is reduced considerably by the uplift volume, and safeguard allowances. In all these cases the effective relief for decommissioning is still at a very high rate.

There is another group of fields where effective decommissioning relief is around 40% and the tax take is in the 52%-62% range. These fields are medium sized or small and often relatively high cost. The significant difference between the tax take and decommissioning relief results from the decommissioning losses for PRT displacing the volume allowance (and occasionally the safeguard benefit). When the decommissioning losses are clawed back they enter periods when the volume allowance is already being utilised. As noted above the PRT rules state that the decommissioning losses should be utilised before the volume allowance. There may not be sufficient income for PRT to absorb both the decommissioning loss clawed back and the volume allowance. As the volume allowance cannot be transferred to an earlier period the value of the allowance is lost. The decommissioning losses have displaced the volume allowance. This is arguably anomalous, especially in fields where in earlier years substantial amounts of PRT may have been payable. This is the main factor which produces the result that the tax take becomes significantly greater than the effective decommissioning relief.

On future fields the tax take is simply the CT at 33%. Looking ahead for some years the attainment of effective relief will depend on the availability of sufficient taxable income for CT purposes. It is possible that some investors may not have sufficient current income for CT purposes to cover their decommissioning costs and may have to utilise the carry back provisions. The three-year carry back provision may not always be sufficient, especially when the investor is decommissioning his "last" field.

PRT Tax Yield and PRT Relief for Decommissioning Through Time

The financial modelling permits a comparison to be made through time of the tax takes and cost of tax reliefs for the expenditures incurred in conducting the decommissioning work. This issue is particularly apposite with respect to PRT. In Charts 40-42 the yield of PRT (excluding the cost of relief for decommissioning expenditure) is shown with the value of PRT reliefs in the period from 1996 to 2015. The decommissioning of the fields is considered on a field stand-alone basis. It is seen that the PRT yield exceeds the cost of decommissioning relief for PRT until well into the next century. Under the $18 price (Chart 40) the yield greatly exceeds the cost of the relief until 2012. In subsequent years the relief exceeds the revenue raised. Under the low price (Chart 41) the relief first starts to exceed the revenue in 2011. Under the high price (Chart 42) the relief first exceeds the revenues in 2014. If the decommissioning were considered on a field interdependent basis the times when the PRT reliefs exceeds the PRT yield would be further postponed.

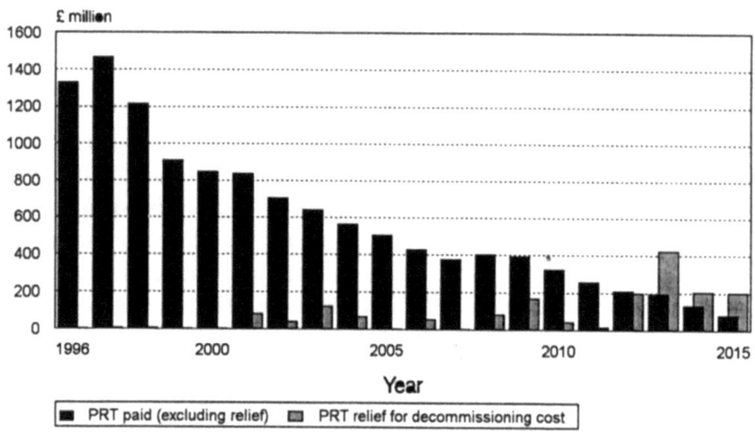

Chart 40: Real PRT and PRT reliefs for decommissioning cost (1996)
(All fields. Stand-alone basis. Base price)

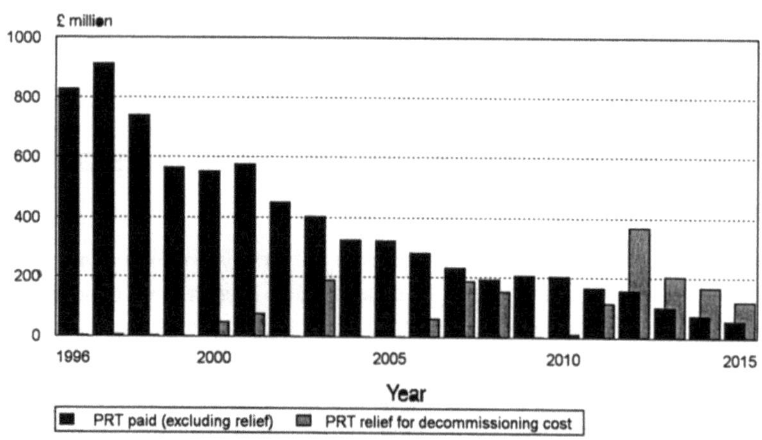

Chart 41: Real PRT and PRT reliefs for decommissioning cost (1996)
(All fields. Stand-alone basis. Low price)

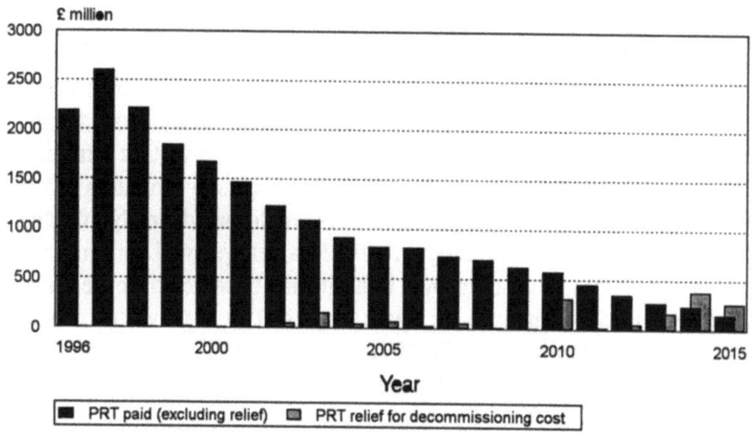

Chart 42: Real PRT and PRT reliefs for decommissioning cost (1996) (All fields. Stand-alone basis. High price)

The Problem of Financial Security and Field Decommissioning

The financial security problem relating to decommissioning has been the subject of much debate. It was highlighted in 1986 following the collapse of oil prices. This raised the possibility that some investors would not be able to fulfil their decommissioning obligations. The UK Government took action to protect its position. It passed the Petroleum Act of 1987 which in essence confers joint and several liability on all licence partners in a group. It appears that this was a new development in some cases, though in other joint venture agreements some elements of joint and several liability were already in existence. The Petroleum Act 1987, also gives the Secretary of State powers to pursue licence partners through their corporate tree in pursuit of assets which may be employed to finance decommissioning programmes.

The Petroleum Act, 1987, was passed against a legislative background which included the Insolvency Act. This has tightened the regulations relating to trading when insolvent. With the possibilities of charges of wrongful trading being placed on directors this Act probably increases the chances of companies in financial difficulties going into liquidation.

Against this background investors may feel that their risk exposure approaches is high. They may wish to procure financial security. This can be obtained by several devises. The advantages and disadvantages of some of the main methods are discussed below:

(a) Third Party Guarantee

This device, normally with banks, could be useful for small companies in particular. The technique has the advantage that the guaranteed amount is limited to the investor's individual share of the decommissioning costs. A bank will require security for providing a guarantee, however, and investors who have only one field may have difficulties in providing the necessary collateral. A further problem is that an investor's borrowing powers would be adversely affected. As a field approaches its decommissioning date the fees charged will certainly increase. When many guarantees are sought by a bank it may become concerned at its own risk exposure with consequential increases in fees. The net cost to investors has been reduced as a consequence of the 1991 Finance Act. The reduction in net cost is, of course, much greater for a PRT payer.

(b) Parent Company Guarantee

In situations where a licensee has a financially strong parent this technique could be attractive. Problems arise where the parent is not financially strong. Co-licensees may then be unimpressed with the guarantee. The existence of guarantees can also adversely affect the parent company's borrowing powers.

(c) Trust Fund

This scheme offers the prospect of providing (a) security to both Government and investors, and (b) effective finance for the decommissioning programme. To procure security the funds need to be alienated. Investors will then wish the contributions to be tax deductible. If they are made on a gross basis the effective cost is much higher if the funds are alienated. The rules of the Fund need to clearly specify how much has to be set aside and over what period. Thus estimates of the decommissioning costs and the remaining field revenues and costs are required. Periodically, these can, of course, be revised. The issue of how the income from the Fund should be treated for tax purposes has also to be clarified. Small oil companies are particularly keen to see the establishment of Trust Funds with tax-deductible contributions. This route of financing decommissioning obligations may be particularly suitable for such investors. As noted above, parent company guarantees may provide insufficient comfort to partners and bank guarantees may become very expensive. The prospect of default by a large co-licensee could also be financially catastrophic for a small investor.

The UK Government has not agreed to the proposal by the industry that a Trust Fund with tax deductible contributions be established. There is concern about over provision of funds and thus over generous tax relief.

In the UK tax relief for an expenditure is generally not permitted until the expenditure has been incurred. The Government is thus concerned that introduction of relief for contributions would lead to inequity among different groups of taxpayers.

There are mechanisms for meeting these objections. The value of contributions on which tax relief is to be obtained could be defined in present value terms in relation to the expected decommissioning date. This could achieve equitable tax treatment among different taxpayers. Any incentive to overprovide could be reduced or entirely removed by adopting two procedures which are employed in the Netherlands where such provisions are tax deductible Their Government has the ability periodically to obtain independent estimates of decommissioning costs. The annual provisions can then be reassessed. If there is overprovision when the decommissioning occurs the Government subjects the overprovision to taxation. In this case the overprovision is subjected to both income tax and State Profit Share (the Dutch equivalent of PRT).

(d) Mutual Guarantee Fund

This scheme would underwrite the failure of a member to meet his decommissioning obligations. It is probable that, if it were to be attractive to licensees, the present legislation dealing with joint and several liability in the UK would have to be modified. Under the present rules a licensee would be accepting further risks without reducing his liabilities. In practice a scheme would probably involve contributions based on likely decommissioning costs rather than the possibility of investor default. One problem is that large investors may feel that they are subsidising small ones.

(e) Government Grant

This scheme could replace the system of tax reliefs for the decommissioning expenditures. For example, it could follow the example of the Norwegian system, where the Government pays a share of the decommissioning costs equal to the tax take over the life of the field in question. Investors would have to find their own share of the costs and the problem of possible default and its consequences emerges again. The question of precisely how tax take over the life of a field is defined also needs to be resolved

(f) Decommissioning Levy

This scheme involves the introduction of a special decommissioning Levy. It would be tax deductible and would procure sufficient revenue to pay for the decommissioning programmes. Government would then assume responsibility for the decommissioning programme. Governments have not shown any interest in this concept, and so it is most unlikely to be introduced.

(g) Insurance

Several possibilities exist here. For example an endowment fund could be established to meet future decommissioning obligations. The insurance premium could be tax deductible, at least in part. Another possibility is that insurance could be taken out by an investor against default by a co-licensee for his share of the costs. It is likely that the premium would be very expensive.

Residual Liability

Under current legislation investors would retain ownership of any remaining stumps of platforms and pipelines under an approved decommissioning programme. They would thus have residual liability against claims arising from such ownership. Investors would also have to undertake any necessary inspection and maintenance work of the remains of installations. The residual liability is in perpetuity.

Investors in the North sea are concerned about accepting residual liability in perpetuity. North Sea oil has a limited life span. The companies, it is argued, have a correspondingly limited life, at least as oil producers. Because of this, and given the expected perpetual life of governments, the industry has felt that the UK Government should take-over ownership of the remains of installations and assume any associated residual liability. The industry has also proposed to make one-off payments into a fund which would meet any claims emanating from residual liability. The intention is that the resources of the fund financed by the industry would be sufficiently large to ensure that the taxpayer did not incur any net costs from assuming ownership.

The UK Government is unconvinced that it should take over ownership of any remains. Residual liability does, of course, exist among property owners throughout the economy, where there is no suggestion that the Government should assume residual liability when a property is no longer in productive use. The question is whether the circumstances offshore are sufficiently different to warrant special treatment. It is certainly the case that many oil companies have been set up to exploit North Sea oil, and, when these activities are exhausted, the raison d'être for these companies disappears. Maintaining a corporate structure to meet any residual liability in practice emerges from the existence of partial removal, and it could then be argued that investors should accept the consequences.

The argument becomes very much finer when the willingness of the industry to finance or buy out the residual liability consequences is taken into account. The legislative and administrative effort required to establish and operate an efficient working scheme then arises as an issue. The comparative novelty of any scheme would entail some innovations and probably changes to the inheritance tax law.

There are other potential solutions to the problem. An entirely private market possibility would involve the oil companies taking out insurance policies for residual liability with one-off premiums being paid. From the viewpoint of the insurance industry the notion of the acceptance of risks in perpetuity by accepting one-off premiums is very novel. It is not clear how willing the insurance industry is to provide such a facility, though presumably if the premiums were sufficiently high the service would be forthcoming.

Organisations which already exist and which provide cover for residual liabilities are Protection and Indemnity clubs. These have existed for a long time in the shipping industry. The P&I clubs are formed by ship owners. They are in effect mutual insurance partnerships, providing for the insurance of third party liabilities and administered by the P&I club. An initial call is payable by each member (in the case of shipping related to the tonnage of each member). The call is estimated to be sufficient to meet all likely future claims. The managers arrange reinsurance. The fund would generate investment income which should be sufficient to cover ongoing administration costs and the continued purchase of reinsurance.

There is obviously some merit in a P&I Club for handling the residual liability problem. An alternative is for an individual investor to make his own insurance arrangements either by one-off or annual payments. Large companies may prefer this. The taxation position regarding one-off contributions requires modifications. At present such payments are regarded as non-allowable capital expenditures and so relief is not available for corporation tax. There is also no relief for PRT. For annual payments relief is available for corporation tax but not for PRT or royalty.

An investor with residual liability relating to the remains of decommissioned facilities will issue some maintenance and inspection costs. These are generally tax deductible for PRT and CT. When a licence expires, however, the relevant PRT terminates as well. Thus, according to present legislation, the investor would not be able to obtain PRT relief for the expenditures incurred following termination of the relevant licence. This is arguably anomalous.

Decommissioning Aspects of Re-Use of Facilities

Given the high cost of decommissioning, the cost of building new structures and the environmental concerns raised by the decommissioning of existing structures on depleted oil fields, there is a case for the re-use rather than the break up of existing structures. This could obviously be the case with a floating platform or FPSO. This section discusses the decommissioning aspects of re-use, although it is recognised that many other aspects will influence any decision.

An option to re-use a depleted field's assets can complicate the cessation of production date decision. Re-use will change both the remaining NPV of the first field and the NPV of the second field which re-uses the structure. If a facility is to be re-used, how it is transferred to the new user has significant cashflow and tax implications for the first field. One possibility is that the investor will use the floating platform on another field belonging to himself. A much more likely possibility is that the asset is sold to a new user. In that event the first field owner would have to pay tax (PRT and CT) on the disposal receipt. The PRT rules also state that if the sale of the asset took place more than two years after it was last used, there would be no disposal receipt for that tax. Where the asset is sold the first field investor's NPV will increase, if as is most likely, the sales value less the taxation on the disposal receipt is greater than the decommissioning cost less decommissioning cost relief and less any PRT on excess expenditure relief. For PRT purposes the expenditure reliefs for the asset are apportioned between the old and new fields according to respective reserves or lives of fields. This happens even if the second field is non-taxable for PRT purposes. There is thus some excess expenditure relief attributed in the past to the old field and this is subject to PRT.

If the asset were re-used on a second field belonging to the same investor, there is again an apportionment of the expenditure relief, and, if the second field is non-taxable, the excess would be subject to PRT. The decommissioning cost is delayed, and, if the asset is classified as a mobile asset, it will be shared between the two fields on a "just and reasonable basis", which could again be the respective reserves or lives of the two fields. Thus, with respect to the first field, the investor is now liable for only a proportion of the final decommissioning cost, and is not liable for this cost until later. The remaining NPV on the old field will fall to a lesser extent if the present value of the attributable real decommissioning cost less tax relief, plus the real PV of the extra taxation due on the excess expenditure allowance, is less than the real present value of the full decommissioning cost less tax relief. These values are, of course, negative from the investor's viewpoint. The tax liability will increase in present value terms from the delay in decommissioning relief that results from re-use, and from the removal of PRT relief on a proportion of the final decommissioning cost. The new field would be subject to CT only, and therefore no PRT relief for the decommissioning cost would be allowed for its proportion of the decommissioning cost.

The issues discussed above were analysed by financial simulation. Fields 1 and 2 are model fields with respective recoverable reserves of 175 and 230 million barrels. They were developed in the 1980's using floating platforms. They are both due to cease production at time **t**. Field 2 has development costs around 4 times as great as Field 1. Both fields pay PRT and CT, but Field 1's PRT liability is low, and it will not have used all its volume allowance by the cessation date. The decommissioning cost of Field 1 at £13 million in real terms is much lower than that of Field 2 at £30 million. The remaining NPVs on Fields 1 and 2 with 1996 as the base year were £85 million and £32 million respectively on the assumption that they were decommissioned upon cessation of production. The effects of re-use of the platform depend upon the characteristics of the new user fields. In the analysis 5 new fields with representative but different cost structures and production profiles were considered. They were all assumed to be available for development at time t.

The relative sizes of the new fields in relation to the size of the old ones are relevant to the determination of the tax effects of re-use and decommissioning. The relationships are shown below in Table 1.

New Fields	*Field 1* (%)	*Field 2* (%)
A	20	15
B	30	25
C	15	10
D	20	15
E	30	25

Table 1: Size of new fields in relation to old fields

On the assumption that the investor redeploys the asset on another field belonging to himself, the remaining NPVs of Fields 1 and 2 from base year 1996 were calculated. The differences in the remaining NPVs including and excluding the facility re-use were also calculated. The results are shown in Chart 43. It is seen that the changes are negative in all cases, meaning that the remaining returns on the two old fields are reduced as a consequence of the re-use. It is also seen from Chart 43 that the decrease in the remaining NPV of Field 2 is greater than that of Field 1. This is because the change in the present value of the tax take is greater with Field 2. The results of this calculation are shown in Chart 44. The increase in taxation arises to a large extent from the extra PRT charged on the excess expenditure allowance. This alone was found to be sufficiently strong to cause the remaining NPVs to be (just) less than they would have been if the old fields had been decommissioned in the orthodox way. The PRT clawback was found to be larger on Field 2 because its development costs are higher than those of Field 1. The real PV of the tax take also increases because of the loss of some PRT relief for the decommissioning cost due to the apportionment of these costs between the old and new fields. On Field 1 there was no PRT relief for decommissioning in the case without re-use and thus no extra penalty occurs when re-use is effected. The increase in the PV of the tax take is again higher on Field 2 because the decommissioning cost is larger on this field. The real PV of the tax take also increases because of the change in the timing of the decommissioning relief which is given. It is given only when the new field ceases production. This is much further in the future, and so the PV of the relief declines. The interest and the associated cap on PRT repayments do not provide adequate compensation for the delay. It is seen from Chart 44 that the increase in taxation is generally higher when the structure is re-used on new Fields B and E. From Table 1 it is seen that these are the largest of the new fields in relation to the old ones. Thus the PRT clawback of excess expenditure allowances is greatest in these cases. It is also seen that the increase in taxation is lowest when the platform is redeployed on new Field C because the PRT clawback of the excess expenditure allowance is smallest in this case.

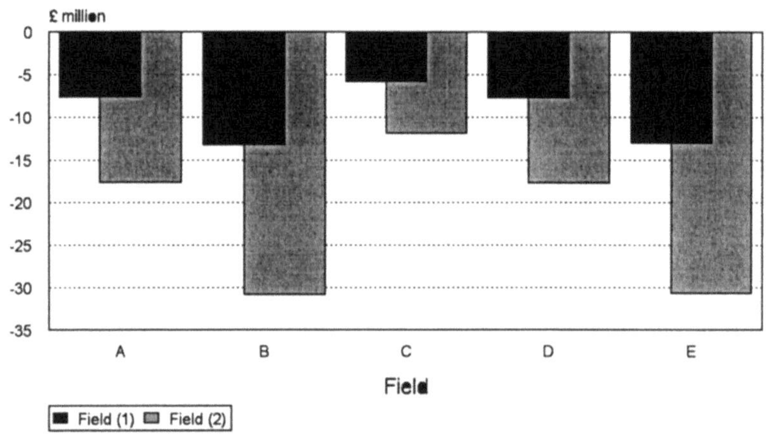

Chart 43: Change in real NPV at 10% of old fields (Asset re-used in investor's own field)

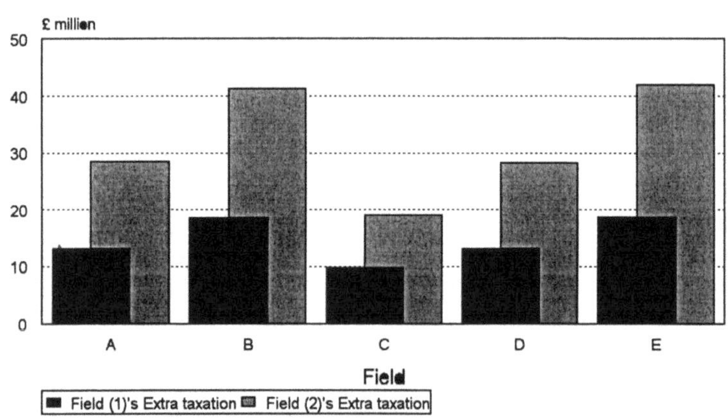

Chart 44: Change in real PV tax-bill at 10% of old fields (Asset re-used in investor's own field)

The case where the asset is sold by the investor in the old fields to a separate investor on the new fields is now examined. The changes in the remaining NPVs on the old fields are shown in Chart 45, and the corresponding changes in the present values of the tax takes are shown in Chart 46. *The value of the sale of the asset is excluded* as the purpose is to highlight the decommissioning aspects. The results indicate the size of the net of tax value from the asset sale which would be necessary to ensure that the investor in the old fields did not lose from the transaction.

It is seen that the pattern of the results is very similar to those already discussed. The declines in the remaining NPVs and the increases in the present values of the tax takes are very slightly larger than in the earlier case. The exercise indicates the complexities of the calculations that the investor in the old fields has to undertake to determine to what extend he really gains from selling the asset.

It should be emphasised that the above analysis has only considered the decommissioning aspects of the re-use of assets. The benefits from re-use depend on many other factors, such as the reduction in the cost of facilities in the new fields. An analysis of this is beyond the scope of the present chapter.

Conclusions

In this study the economic, financial and fiscal issues relating to field decommissioning the UKCS have been examined. The findings may be summarised as follows:

1. Three criteria for determining the optimal timing of field abandonment have been discussed namely (a) negative net profits, (b) minimum return on ongoing expenditures, and (c) maximisation of remaining net present value. The last is the most scientific as it takes into account all the relevant factors including future oil prices and decommissioning costs. The relevant discount rate may be lower than that employed for assessing the field when it was initially being considered for development, but the risks are still substantial at the decommissioning decision time.

2. Employment of the remaining net present value criterion very frequently leads to the optimal timing of the decommissioning coinciding with that under the net profits criterion. In a number of cases, however, it may be in the interests of the investor to postpone decommissioning beyond the point when profits turn negative. The larger the decommissioning costs and the higher the discount rate employed the greater is this incentive. In a few cases the incentive to postpone field decommissioning beyond the introduction of negative net profits could continue for a considerable number of years. The existence of tax relief on the production and decommissioning losses plays some role in the determination of the optimal decommissioning date. It was

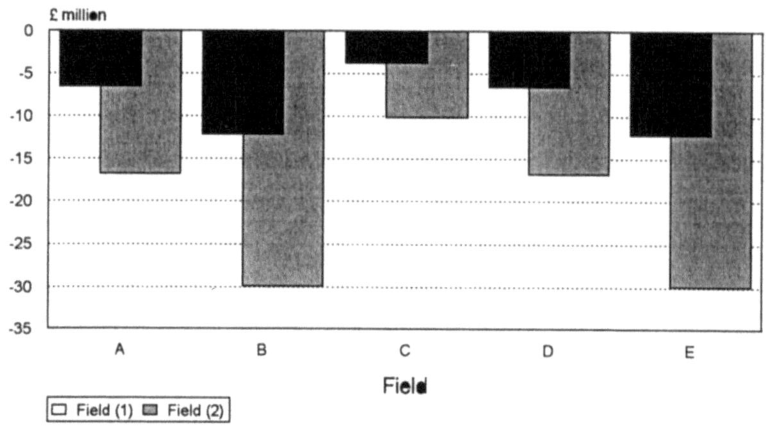

Chart 45: Change in real NPV at 10% of old fields (Asset sold by first field investor. Sales value not included)

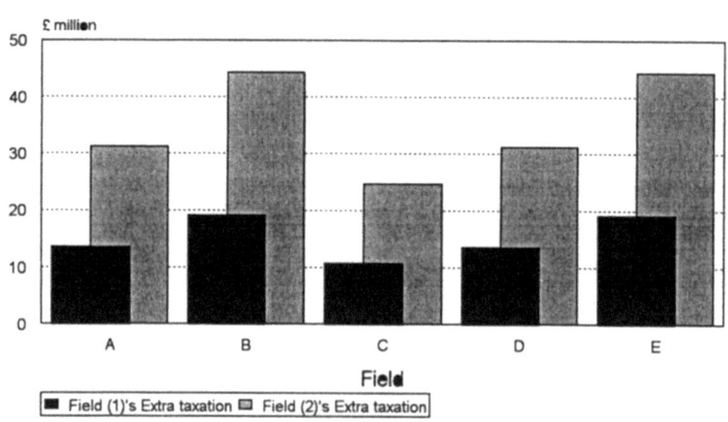

Chart 46: Change in real PV tax-bill at 10% of old fields (Asset sold by first field investor. Sales value not included)

found that, with high discount rates and large decommissioning costs, when remaining NPVs on a pre-tax basis were maximised some time after negative profits had emerged, the optimal time on a post-tax basis was accelerated somewhat. The tax reliefs did not fully compensate for the delay in decommissioning which had been justified on a pre-tax basis.

3. Investors may sometimes wish to defer decommissioning of installations because they have prospective uses for the facilities at some time in the future beyond the point when the field to which the installation relates is economically depleted. In such cases cessation of production and mothballing of facilities may be considered. Employing the remaining net present value criterion for determining field decommissioning was sometimes found to be consistent with mothballing of facilities for long periods, though more often mothballing resulted in a reduction in remaining NPVs. Again the size of the discount date and the decommissioning costs played key roles in determining the outcome. Tax relief for the mothballing costs was found to be a significant factor in the determination of the optimal timing. It should be noted that such tax relief is allowed only when the mothballing is approved by the Government.

4. The pattern and timing of the numbers of fields being decommissioned over the period to 2015 was found by financial simulation. Future real oil prices of \$23, \$18 and \$15 were used. Over the next decade there is likely to be a gradual build-up in the number of fields decommissioning in the UKCS. The numbers will increase significantly from around 2001 or so, and they become even greater in the period 2006-2010. In the latter period 20 fields or more could become decommissioned yearly, creating a potential problem of inadequate capacity and consequential cost increases in the relevant contracting activities. The fact that some major facilities will be employed to process and transport petroleum from other fields after the original field to which the facility related is depleted will postpone the decommissioning activity to some extent. A significant oil price sensitivity to the timing of the decommissioning activity was also found. It is to be expected that (a) technological progress, (b) cost reductions, and (c) discovery of new reserves, will postpone the decommissioning of some fields beyond the dates shown in the findings of this study.

5. The complex system of tax reliefs for decommissioning expenditures in the UK operates in an erratic manner. In around 68% of fields the decommissioning relief was at a lower rate than the overall effective tax rate. In many cases the difference was slight, however, reflecting the lack of relief for royalty in the Southern basin. On a few larger fields decommissioning relief can exceed the effective tax take. This follows when decommissioning relief at the full rate is obtained but the PRT take on the field benefits from the uplift, volume and safeguard allowances. On a significant number of medium and small sized fields decommissioning relief was in the 33%-40%

range while the effective tax takes were considerably less in the 40%-60% range. This is caused by the decommissioning losses displacing other allowances for PRT, especially the volume allowance.

6. The cost to the Exchequer of decommissioning tax relief is not going to be large for some years. Many of the facilities which are likely to be decommissioned in the next few years are typically not very expensive. The cost of PRT tax reliefs for decommissioning should not exceed the revenues from PRT until 2012 and thereafter. This finding broadly applies under all the 3 oil price scenarios.

7. The issue of financial security and decommissioning obligations may be addressed in several ways. The UK Government has protected itself from the possibility that decommissioning obligations will not be undertaken by imposing joint and several liability in the Petroleum Act, 1987. Investors are thus encouraged to find devices to pursue financial security for themselves. This can be obtained in several ways including (a) bank guarantees, (b) parent company guarantees, (c) mutual guarantee fund, (d) government grant, (e) decommissioning levy, (f) insurance, and (g) Decommissioning (Trust) Fund. There are attractions and problems with all of these devices. The concept of an Decommis-sioning Fund is attractive to investors when the contributions are tax deductible, but costly when contributions have to made on a gross basis. The UK government has rejected the idea of a fund with tax deductible contributions. There is apparently concern about overprovision and inequitable tax treatment among taxpayers. In the UK tax relief is generally not given for an expenditure until it is incurred. These objections can be met by suitable design of the scheme. Thus the contributions eligible for tax relief could reflect the present value of the expected decommissioning costs. This should procure equity among taxpayers. The Government can take powers to have a periodic independent assessment of the likely decommissioning costs. If overprovision is still eventually made, the sums in question can be subject to corporation tax and PRT.

8. The issue of residual liability for third party claims following field decommissioning may also be addressed in several ways. One possibility is for the Government to take over this role. The UK Government has rejected this idea, even with the promise of the availability of a fund financed by the industry sufficient to meet likely claims. Other private solutions are possible including the formation of Protection and indemnity Clubs and individual insurance schemes by investors. The taxation arrangements with respect to such schemes require attention, especially those involving one-off contributions. At present such payments are regarded as non-allowable capital expenditures and so relief is not available for corporation tax. There is also no relief for PRT. For annual payments relief is available for corporation tax but not for royalty.

9. An investor with residual liability relating to the remains of decommissioned facilities will have some maintenance and inspection costs. These are generally tax deductible for PRT and CT. When a licence expires, however, the relevant PRT terminates as well. Thus, according to present legislation, the investor would not be able to obtain PRT relief for the expenditures incurred following termination of the relevant licence. This is arguably anomalous.

10. The possibility of re-use of some facilities such as floating platforms is currently a live issue. The decommissioning aspects of this activity were examined. So far as the investor in the old field is concerned his position with respect to the decommissioning aspects to a significant extent depends upon the tax treatment of the related activities. If he sells the asset to another party, the sales value is regarded as a disposal receipt for PRT. If, however, the asset has not been used for two years prior to the sale it is ignored for PRT purposes. For PRT purposes if the asset were used on two fields and was classified as a mobile asset, only a proportion of the overall expenditure reliefs are attributable to the old field. This happens even when the new field is not subject to PRT. The proportion depends on the relative sizes or length of lives of the old and new fields. Any excess expenditure allowance is subject to PRT.

    In financial simulations of the position facing the investor in the old field it was found that the PRT clawback of expenditure reliefs constituted a significant negative element of the investor's position at the time of decommissioning. In the financial simulations of the old fields the remaining NPVs were positive but the PRT clawback of the excess expenditure allowances was sufficiently strong to reduce them. There was found to be a high degree of sensitivity in the investor's financial position on the old field to the relative size of the new field on which the asset was to be deployed. The small the new field in relation to the old one the smaller is the PRT clawback of the excess expenditure allowances.

    In the situation where the investor used the asset on a new field of his own there is a further complication regarding relief for the decommissioning costs. The new field will not be subject to PRT. When the asset is decommissioned at the end of the life of that field a proportion of the costs involved can be set against PRT on the old field. This proportion depends on the respective sizes or lengths of lives of the two fields. In the financial simulations it was again found that the PRT payments on excess expenditure allowances constituted a significant negative item with respect to the remaining NPVs on the old fields.

5

Legal Aspects of Decommissioning Offshore Structures

A D M Forte

Introduction

This chapter is intended to provide readers, especially those who are not lawyers, with an outline of some aspects of the legal regimes applicable to the decommissioning of offshore oil installations. While it covers a fairly wide area, some specialised aspects of decommissioning, such as taxation, are beyond the writer's competence and have been omitted for that reason.

The Offshore Environment: The Interface of Law and Technology

International law confers on a coastal state sovereign rights in respect of the mineral resources found on and under its continental shelf. This means that such states have the exclusive right to exploit these resources [1]. But coastal states do not enjoy unlimited rights in this respect. In particular, installations or devices for the exploration and exploitation of mineral resources, and any safety zones placed around them, cannot be located in a position which interferes with essential shipping routes or sea lanes [2]. What constitutes a state's continental shelf, however, depends on the person to whom the question is put [3]. For lawyers the answer depends on the current state of international law regarding the seabed adjacent to a state's shores.

Article 1 of the 1958 United Nations Convention on the Continental Shelf (Geneva Convention 1958) conceived of the continental shelf as comprising the seabed and its subsoil, measured from where its territorial sea ended, to a depth of 200 metres in all cases and beyond that depth where exploitation was possible. On a literal interpretation of this statement, a state's continental shelf would creep relentlessly forward as offshore technology advances until even the ocean floor falls within its confines. In the 1982 United Nations Law of the Sea Convention (UNLOSC 1982) distance is substituted for depth and the continental shelf is defined as stretching to either the outer limit of the adjacent continental margin or, where that margin does not extend 200 nautical miles from the baselines from which the breadth of the territorial sea is measured, then to that distance [4]. The International Court of Justice in the *Libya/Malta Continental Shelf Case* [5] was of the opinion that Article 76 of UNLOSC 1982 reflects a principle of customary international law that all coastal states were entitled to an 'Exclusive Economic Zone' of 200 nautical miles extending seawards from the baseline from which the breadth of the territorial sea is measured [6].

Offshore oil and gas operations are conducted at varying distances from the coastline and in varying depths of water. In the case of the United Kingdom, exploration for and exploitation of mineral and gas resources up to 400 miles from land and in waters more than 1,000 feet have been authorised [7] and there are currently something in the region of 150 production platforms operating on the UK's continental shelf (UKCS). These operations are supported by a variety of other installations and there is a network of around 5000 miles of pipelines in place on the seabed of the UKCS [8]. Authorised by the Government of the UK, these operations are subject to the jurisdiction of the Scottish and English courts [9]. But just as (a) the authority to permit extraction of offshore oil and gas reserves is conferred on the UK by international law, and (b) extraction operations themselves are permitted and controlled by its Government, so too is the issue of what happens to offshore installations when a field's reserves have been exhausted. Indeed, permanent offshore installations cannot be sited on the UKCS without the Department of Transport's consent, and it is normally a term of any such consent that abandonment of the site will be effected in a way which will ensure that no danger or obstruction to navigation is likely to occur [10]. In addition, Model clause 19(3), found in production licences, confers on the Secretary of State the power to require the removal of platforms [11]; and the works authorisation [12] for the laying of pipelines in the territorial sea or UKCS may require their removal after their use has ended.

Decommissioning or Abandonment: What's in a Name?

Decommissioning and abandonment are terms which describe 'the process undertaken by the operator of an oil or natural gas installation to plan, gain approval for and implement the removal, disposal or reuse of an installation when it is no longer needed [13]. So far as actual disposal of platform installations is concerned this may currently be achieved in several ways, for example: (1) retention on site but conversion to another use (*e.g.*, as an artificial reef); (2) 'mothballing' on site; (3) relocation for use on another site; (4) emplacement or toppling on site; (5) removal to land for disposal on a licensed waste-disposal site; (6) removal to land for recycling; and (7) disposal in deep water. For pipelines there are 3 options: (1) clean them and leave them in place; (2) bury them in place; (3) completely remove them.

Geneva Convention 1958, Article 5.5, refers to the abandonment of offshore installations. UNLOSC 1982, Article 60.3, does the same. Part I of the Petroleum Act 1987 is headed 'Abandonment of Offshore Installations' and refers, throughout, to the disuse of such installations as 'abandonment'. The term is also used in some of the provisions of the International Maritime Organisation's (IMO)1989 *Guidelines and Standards for the Removal of Offshore Installations and Structures on the Continental Shelf and in the Exclusive Economic Zone* [14]. In the oil industry itself and in related professions, including the legal profession, 'abandonment' is commonly used. And yet, despite its widespread use, 'abandonment' is an unfortunate choice of word and 'decommissioning' a preferable substitute to describe the process and procedures associated with disposals of installations after they are no longer needed.

If one places an item in a dustbin or refuse site one abandons it and, under Scots Law, ownership ceases. But the item does not become ownerless, for abandoned property belongs to the Crown in accordance with the principle '*quod nullius est fit domini regis*' [15]. And abandonment of an item does not preclude the abandoners from being liable to make reparation in damages should they injure another in circumstances where the possibility of their causing harm was foreseeable [16]. Under English Law it appears that property cannot be abandoned easily and nor can one escape liability for harm caused by property by saying that it has been abandoned [17]. These statements are broadly true of the abandonment of offshore installations. But when an offshore operator declares an intention to abandon an installation, it does not matter that under both international and domestic law 'abandonment' is linked with *removal*, nor that an abandonment programme has to be prepared, submitted to and approved by the Secretary of State for Energy, nor that the operator may be liable to make reparation to vessels or other equipment damaged by coming into contact with what has been left on the sea bed, nor for that matter that the operator has authorised complete removal, the layman, relying on newspaper or television coverage of events, sees, or may be persuaded to see, abandonment as walking away from environmental responsibility. In the case of Brent Spar it did not matter that the science and engineering were correct in asserting that deep ocean disposal of a substantially cleaned installation was the best option, eco-hype went into overdrive and even the *Financial Times* ran the headline 'Shell vows to stick to dump plan' [18]. Abandonment suggests, or can be made to suggest that an offshore operator is entitled to walk away from any installation once it has served its purpose. But it is worthwhile to recollect that of the seven disposal options referred to above, only two, emplacement and deep-sea disposal, permit one to walk away and then only after strict compliance with the abandonment provisions in Part I of the Petroleum Act 1987.

The International Decommissioning Regime

In relation to decommissioning, it is the avowed object of the Government of the UK to comply with 'the requirements of international law' [19]. The Government has also reiterated its adherence to the IMO *Guidelines and Standards* and has even extended these on occasion [20]. Compliance with international law is important, since a state which is in breach of its international obligations is liable to make reparation for damage suffered by another state in consequence of that breach [21]. This section will describe the applicable rules of international law governing the decommissioning of offshore installations on the UKCS

(a) Offshore Installations

At the outset one has to be clear what is comprehended by the terms 'installation' or 'structure' which occur in the various conventions. Basically the problem arises in connection with pipelines. If these are installations or structures, then the UK must observe international law when providing for their disposal. If they are not, no such obligation is laid on the UK. It is clear that offshore production platforms, well heads, storage and loading units are 'installations' or 'structures'.

Geneva Convention 1958, Article 4, confers on states an entitlement to lay submarine pipelines on the continental shelf. Article 5.2 confers an entitlement to construct and operate installations on the continental shelf. From this separate treatment, of pipelines on the one hand, and installations on the other, it is possible to infer that the reference to installations was not intended to comprehend pipelines. However, as Article 5.2 refers not only to 'installations' but also to 'other devices necessary for ... the exploitation of' the natural resources of the continental shelf, it is certainly arguable that pipelines in general, and those between production platforms and shore facilities in particular, are just such 'devices'. Against this argument, however, is Article 5.5 which only refers to an obligation to remove abandoned or defunct 'installations': which could not, of course, include pipelines. The better view, on balance, would be to state that pipelines are not covered by the decommissioning provision of the Convention. UNLOSC 1982, while referring to 'installations' and 'structures', defines neither term and it seems unlikely that it applies to pipelines either [22]. The IMO *Guidelines and Standards* do not apply to pipelines and neither the Oslo nor London Dumping Conventions, which will be discussed later, is thought to apply to pipelines.

While the decommissioning of pipelines under international law appears to be free from restraint, the UK does, nonetheless, regulate this process under the Petroleum and Submarine Pipelines Act 1975, the Food and Environment Protection Act 1985, and the Petroleum Act 1987. The matter will be considered again when we examine the UK's domestic regime for decommissioning.

(b) Removal Options

Geneva Convention 1958, Article 5.5, stipulates in the clearest language that decommissioned installations must be entirely removed from their locations. Disposal on land was, therefore, the only option available. In contrast, Article 60.3 of UNLOSC 1982, although it refers to the removal of offshore installations, specifically sanctions partial removal. The text of the provision reads: 'Due notice must be given of the construction of such artificial islands, installations or structures and permanent means for giving warning of their presence must be maintained. Any installations or structures which are abandoned or disused shall be removed to ensure safety of navigation, taking into account any generally accepted international standards established by the competent international organisation. Such removal shall also have due regard to fishing, the protection of the marine environment and the rights and duties of other States. Appropriate publicity should be given to the depth, position and dimensions of any installations or structures not entirely removed.' Although not a party to the 1982 Convention, and, therefore, bound only by the 1958 Convention with its strict requirement of complete removal, the UK Government takes the view that UNLOSC 1982, Article 60.3, represents the customary rule of international law and that Article 5.5 of the 1958 Convention has to be read in a way which is consistent with this customary rule. Furthermore, although the Government does not actually say that Article 5.5 is no longer relevant to an age when installations such as Gullfaks C and Troll respectively weigh in at 1.3 and 3.25 million tonnes (tonnages which could scarcely have been imagined in 1958), it comes close to adducing this as another reason for following the law as reflected in the 1982 Convention [23]. The partial removal of installations is also sanctioned, in appropriate circumstances, by the 1989 IMO *Guidelines and Standards* and it is to these that we now turn [24].

(c) International Decommissioning Standards

The 'competent international organisation' referred to in UNLOSC 1982, Article 60.3, is the IMO: whose *Guidelines and Standards* contains 3 sections of which the first, complementing Article 60.3, imposes a 'general removal requirement' of abandoned or disused offshore installations and structures whether on the continental shelf or in an exclusive economic zone. So complete removal is clearly seen as the ideal. However, the remainder of section 1.1 sanctions both non-removal and partial removal where these options are consistent with the guidelines and standards prescribed. Since it is easy to conclude from this section that it represents a charter for walking away from environmental as well as other responsibilities, it is worth examining in more detail just why such a view would be wrong.

Section 2.1 contains guidelines to assist states in deciding whether to permit either non-removal or partial removal of installations. There is an obligation in such cases to take into consideration the effect this would have on navigation or other uses of the sea (*e.g.*, fishing), and also its effect on the marine environment [25]. Other considerations include the risk that material left on site may shift position, the risk to those who would have to carry out work on an installation if complete removal were required [26], and the possibility of putting an installation to some other use if it were left on site. Finally, cost and technical feasibility are also to be taken into consideration. No decision to permit non-removal or partial removal constitutes a precedent [27]. Each case must be evaluated separately and on its merits. Furthermore, states are free to impose removal requirements which are even more stringent than those prescribed in the *Guidelines and Standards* [28].

Section 3 of the *Guidelines and Standards* prescribes standards which must be applied when deciding which option to sanction. The most important of these is the requirement for the complete removal of all installations below a specified tonnage and standing in a specified depth of water. Any installation emplaced prior to 1 January 1998, which stands in less than 75 metres of water and which weighs (excluding deck and superstructure) less than 4,000 tonnes must be removed entirely. For installations emplaced after that date, the depth criterion rises to 100 metres. And 1 January 1998 marks another important watershed. After that date only installations whose designs make complete removal technically feasible may be sited on a continental shelf or in an exclusive economic zone [29]. The remains of partially removed installations which are sited in waters deeper than the prescribed minima must leave a minimum clearance (a 'water column') of 55 metres above them [30]. If complete or partial retention on site would enhance marine life, or serve some other use, then it may be left in place [31].

(d) Decommissioning and Dumping

The issue here is whether or not the disposal of offshore installations by emplacement or deep-sea disposal amounts to dumping. The Oslo Convention [32] (to which the UK is a party, and which covers the North Sea and parts of the north-east Atlantic and Arctic Oceans) applies to fixed and floating offshore platforms and prohibits the 'disposal' from these of substances such as cadmium and mercury in any decommissioning regime [33]. Other substances, *e.g.*, arsenic, copper, zinc and cyanides, and objects, such as containers and scrap metal, may be dumped if a specific permit to do so is granted by the state sanctioning the decommissioning of a platform [34]. In the case of containers, scrap metal and 'bulky waste', such disposal must be in 'deep water'; which is defined to mean water which is not less than 2000 metres in depth and not less than 150 nautical miles from the nearest land. Whether or not the Oslo Convention applies to the disposal of an offshore platform, as distinct from the dumping of materials from that platform is said to be unclear [35]. However, Article 19.1 of the Convention defines dumping as the 'deliberate disposal of substances and materials into the sea *by or from* [fixed or floating platforms].' The italicised words 'by or into' cannot be interpreted as referring to the item from which materials are dumped. The better view, it is submitted, is that the Oslo Convention does not cover the disposal of platforms. Furthermore, and again paying close regard to the wording of Article 19.1, it can be seen that pipelines are not subject to the Convention. The laying of pipelines along the seabed does not constitute a 'disposal' and their disposal *in situ* after their working life has ended does not constitute disposal '*into* the sea.'

Although the Oslo Convention probably does not apply to the disposal of installations, in 1991 the states who were parties to it adopted *Guidelines for the Disposal of Offshore Installations at Sea*. These OSCOM *Guidelines* are consistent with and complementary to the IMO *Guidelines*. The most important provisions are: (a) the requirement of a specific permit for the disposal of an installation; (b) the development of criteria for disposal on a case by case basis; (c) the permission of disposal at sea where this represents the 'option of least detriment'; and (d) the requirement to take all practical steps to mitigate the impact of disposal at sea on the marine environment. There is also an obligation on states which are party to the Oslo Convention to notify other states parties thereto of any intended disposal at sea of an installation stating, amongst other matters, the reason why such disposal is sanctioned.

The UK is also a party to the 1972 London Convention on the Prevention of Pollution by Dumping of Wastes and other Matter. Unlike the Oslo Convention, this convention defines dumping to include specifically the disposal at sea of 'platforms or other man-made structures' [36]. However, leaving a platform on site or, after partial removal, leaving a stump on the seabed does not, it is thought, constitute a 'disposal'. This is because there can be no disposal '*at* sea' of a platform which is already located on the seabed and which was not, in the first place, sited there in order to dispose of it. It has to be said, however, that there is disagreement amongst lawyers on this point, with some taking the view that disposal of platforms in the manner just described does constitute dumping [37]. However, where the IMO *Guidelines and Standards* apply the point turns out to be rather academic. Whether the London Convention would regard the leaving of pipelines after their use has ended as dumping is unclear. Since the definition of dumping includes the disposal of 'man-made structures', it is arguable that this must include pipelines. But as with platforms, the leaving of these in place after their working-life is over cannot easily be treated as disposal 'at sea'.

In 1992, in Paris, the states parties to the Oslo and London Conventions adopted a new one, the Convention for the Protection of the Marine Environment of the North East Atlantic. This is not yet in force, but when it is it will replace the other two conventions. Under this new convention the status of pipelines is made clear: they are subject to its regime [38]. The siting of offshore structures and pipelines, defined as 'pipelines which have been placed in the maritime area for the purpose of offshore activities', does not constitute dumping. But though 'dumping' is defined to include the 'deliberate disposal in the marine area of offshore installations and offshore pipelines', the decommissioning options of emplacement or toppling and leaving these on location do not constitute dumping provided that 'such operation[s] takes place in accordance with any relevant provision of the Convention and with other relevant international law' [39]. In relation to disposal, states parties are obliged to require that this be done in accordance with 'Best Available Techniques', 'Best Environmental Practice', and the use of 'clean technology'. If 'clean technology' means the utilisation of removable materials only, then, while this can clearly be done in the case of platforms, storage units and drilling vessels, its application to pipelines may be more problematic. However, since economic feasibility is one of the criteria to be taken into consideration, complete removal will not always be necessary.

The UK Decommissioning Regime

When the UK adopts an international law convention, it then enacts legislation to give that convention force of law in the UK. Of the several conventions referred to above, the following have been implemented in this manner:

Convention	*Legislation*
Geneva 1958	Continental Shelf Act 1964
Oslo and London 1972	Food and Environment Protection Act 1985 (Part II)

[40]

The Petroleum Act 1987 was passed in order to permit the UK to fulfil its international obligations regarding the decommissioning of offshore installations. The 1987 Act either changed or clarified the law in several important respect; for example: (a) it confers power on the Secretary of State to require submission to him of abandonment programmes for the proposed removal of both offshore installations and submarine pipelines; (b) it empowers him to make regulations for the 'abandonment' of installations and pipelines; (c) where there are joint operators, it makes each one liable to implement an abandonment programme; and (d) it creates and provides punishments for a number of offences. Where more than one operator has an interest in a platform or pipeline, the 1987 Act plays an important role regarding the drafting of an abandonment agreement.

(a) Abandonment Programmes

Notice calling for an abandonment programme may be given to the operator of a platform or pipeline, the licensee or licensees under a production licence, any party to a joint operating agreement, and even to holding or subsidiary companies of any of these [41]. The abandonment programme itself must do the following: (1) estimate the cost of abandonment; (2) specify the time scale applicable to the proposed abandonment; and (3) where complete removal is not proposed, make provision for continuing maintenance of what is left [42]. This programme may be approved completely, or conditionally or with modifications, or it may be rejected [43]. If an abandonment programme is rejected, the Secretary of State may devise one. He may also do this where, having called for an abandonment programme, none is forthcoming [44]. Where he is not satisfied that there has been compliance with an approved abandonment programme, he may serve a notice ordering remedial action to be taken. In the event of non-compliance with such notice, the

Secretary of State may execute the remedial work required and recover its cost from the defaulting party [45].

The giving of notice requiring the submission of an abandonment programme, the approval or rejection of such a programme, the determination of changes to an approved programme [46], the withdrawal of approval for an abandonment programme [47], and the serving of notice requiring a party to take steps to ensure financial ability to implement an abandonment programme [48], are all actions of the Secretary of State which may be challenged in court under section 14 of the 1987 Act. However, anyone seeking judicial review must do so within 42 days of the action complained of. Furthermore, there are only on two grounds of challenge, namely: (1) that the action was invalid because it exceeded the powers given to the Secretary of State; or (2) that he did not observe any relevant procedural requirements for the exercise of his powers. It is thought, however, that the courts would entertain an action for judicial review based on grounds other than that the Secretary of State acted beyond his powers [49]. If, for example, the decision to take action was based on a misinterpretation of some provision of the 1987 Act, or if he took into account factors which should have played no role in his decision, then, such decisions might be successfully challenged.

(b) Abandonment Regulations

The Secretary of State has the power to make regulations relating to the abandonment of offshore installations and submarine pipelines. Such regulations may prescribe safety standards for their dismantling, removal, or retention *in situ* if not removed completely. They may also provide for the prevention of pollution, inspection and the costs of inspection and payment of fees [50]. To date, however, these powers have not been exercised and the Department of Energy has not promulgated any regulations since, as already noted, it is Government policy to comply with the requirements of international law [51].

(c) Joint and Several Liability

Where responsibility for an abandonment programme is shared among several participants in an installation (*e.g.*, joint licensees) or, in the case of a pipeline, its joint owners or the participants in any tariffing agreement, then, in the words of the Act: 'it shall be the duty of each of the persons who submitted it to secure that it is carried out and that any conditions ... are complied with' [52]. Not only does this have the effect of imposing collective liability on the participants, it also makes each participant individually liable to implement the programme. Lawyers term this 'joint and several' liability. Should a participant default in its duty under an abandonment programme, the others remain liable for the execution of the whole programme [53]. In a doomsday scenario where all of the participants save one have gone into insolvent liquidation, the survivor must bear the cost of implementing the programme alone.

Since the cost of implementing an abandonment programme can be high, the risk of default or liquidation increases the exposure of joint participants. Consequently, the 1987 Act adopts something of a carrot and stick approach to ensure that such participants are financially able to meet their potential liabilities.

Section 10(4) represents the stick. The Secretary of State has the power to require information and documents relating to the financial affairs of any party to whom an abandonment notice has been sent [54]. If he is not satisfied that a party can meet the cost of implementing an abandonment programme, he can order that party to take steps to ensure that implementation can be achieved. The most obvious course of action here would be to order the making of Abandonment Security Agreements (ASA). Although compulsory ASAs do not represent current policy, the possibility of compulsion may still have a coercive effect on industry practice.

The carrot is provided by subsections 3(2) and (3). These subsections provide that notices requiring the submission of abandonment programmes will not be served on banks or companies associated with operators, licensees, owners, etc, if the Secretary of State is satisfied with the financial and other arrangements made to implement the abandonment programme. The practical effect of this is to provide an incentive to make ASAs.

The ASA is often found annexed to a Joint Operating Agreement (JOA). Typically an ASA will provide for 3 things: (1) the preparation of an abandonment programme; (2) the sharing of liabilities; and (3) the provision of security for each party's share of the cost of abandonment JOAs have to be submitted to the Secretary of State for Trade and Industry for approval and an ASA containing the features outlined above will relieve associated companies from having to submit an abandonment programme. The security for a participant's share of abandonment costs may take the form of a guarantee by a parent company, if there is one, and if there is not, from a bank. Where there is no parent company and a bank guarantee is an unattractive option (usually for reasons of costs), then a trust fund, into which contributions are made to fund the cost of abandonment, is the only option.

(d) Offences and Punishments

It is an offence for anyone required to submit an abandonment programme, and who is asked to identify others who may be similarly required, to fail to do so without reasonable excuse [55]. It is an offence to fail to supply the Secretary of State with information relevant to the preparation of an abandonment programme when called on to do so [56]. Where an abandonment programme, or a condition on which approval of the programme was made subject, is not complied with, the Secretary of State may order that action be taken to remedy the default. Failure to comply with such a direction is an offence (unless it can be proved that due diligence was exercised to avoid failure) [57]. Finally, it is an offence for any person to whom an abandonment notice has been given, and who has been required to provide information concerning its financial resources, to fail to do so without reasonable excuse or to provide false information [58]. The penalties for such offences range from fines to imprisonment of a company's directors, secretary, or managers [59].

Decommissioning and Salvage

It is important to establish the ownership of decommissioned items which have been left on site. So far as pipelines are concerned, if these have not been removed, ownership vests in the Secretary of State [60]. Casings and fixtures associated with plugged and sealed wells may also become the property of the Secretary of State on expiry of a production licence [61]. Platforms and floating units, however, as already mentioned, remain the responsibility of their former operators who are still regarded as their owners. That ownership raises the issue of salvage. While Brent Spar was being towed to the site chosen for its disposal, one of the suggestions attributed to Greenpeace was that they might salvage the unit. Can decommissioned installations be salvaged and the salvors claim a reward?

The 1989 London Salvage Convention defines 'salvage operations' as actions taken to 'assist a vessel or any other property in danger' in both navigable and non-navigable waters [62]. This definition does 2 things:

(a) it forms the basis for the identification of the only types of objects which may be salvaged; and

(b) it describes the circumstances in which those objects can be salvaged.

(a) Salvageable Property

Only ships, craft or 'any structure capable of navigation' are classified as vessels. Items whose capability for navigation may be questionable (*e.g.*, floating cranes and semi-submersible lift barges) are capable of being salvaged since they may be classed as 'property' [63]. 'Property' is defined to mean anything which is not 'permanently and intentionally attached to the shoreline' [64]. This would seem to preclude the salvage of any offshore installations which are attached to pipelines leading to the shore or, indeed, of pipelines themselves. However, Article 3 of the London Convention does not entirely exclude the possibility of salvage of these. This provision reads as follows: 'This Convention shall not apply to fixed or floating platforms or to mobile offshore drilling units when such platforms are on location engaged in the exploration, exploitation or production of sea-bed mineral resources.' Clearly, then, any decommissioned offshore 'platform' which is being towed from its location for deep-sea disposal is, potentially, salvageable property. So was Brent Spar a 'platform' for the purposes of the London Convention? No definition of this term is provided by the Convention and, in any dispute, the classification of a structure would depend on the testimony of oil industry experts. That said, it is fairly certain that Brent Spar would be regarded as a platform. Feeder pipelines running from production platforms to storage platforms or loading units, or underwater storage modules themselves are also capable of being salvaged since these may be described as 'property' which is not attached to the shore.

(b) When Are Things Salvageable?

There is a simple answer to this question - when they are in danger [65]. If during the course of transit from location to disposal site, an offshore installation were to break loose from its tugs and be in danger of fetching up on the shore with attendant risks to itself, to navigation and to the marine environment, then it is potentially salvageable. In these circumstances a salvor might even be entitled to an enhanced reward if the installation or its contents 'threatened damage to the environment' [66]. The Brent Spar was in no danger during the course of towage and it may have been injudicious use of the word 'abandonment' which caused misunderstanding. More fundamentally, however, one cannot salvage property when its owner does not wish it to be salvaged. Article 19 of the London Convention addresses the issue: 'Services rendered notwithstanding the express and reasonable prohibition of the owner or master of the vessel or the owner of any other property in danger which is not and has not been on board the vessel shall not give rise to payment under this Convention.' It is true that there is a caveat, the owner's refusal to accept an offer of salvage services must be 'reasonable'. So it is possible that where threat to the environment is present an offer to provide salvage services may not be refused - *provided*, of course that the property posing such threat was in danger. One can only repeat that this was not the case with the Brent Spar.

Residual Liability for Decommissioned Installations

Where offshore installations and pipelines are left in place after having been decommissioned, the issue of liability for damage caused by these presents itself. As the law currently stands, liability is fixed on the owner of the decommissioned installation or pipeline [67]. In their 1991 Report, the House of Commons Energy Committee supported a UKOOA proposal that liability for decommissioned installations should vest in the Government in return for a lump sum payment [68]. The Government has declined acceptance of residual liability, pointing to the 500 metre safety zones around installations and arguing (somewhat hopefully) that the offence of unauthorised entry into these zones will serve as a deterrent [69].

Conclusion

The disposal of offshore installations and pipelines is a closely, although not always clearly, regulated legal process. International law imposes obligations and constraints on the UK and the UK legislates in conformity with these. The hallmark of both the international and national legal regimes governing decommissioning is flexibility in their application; witness the general rule that many decisions are taken on a case by case basis. On the whole the law appears to be appropriate for the state of the technology. That said, the Government possesses some fairly formidable powers with regard to abandonment programmes and with more fields in the North Sea beginning to approach the end of their productive lives, the need for expert legal advice on decommissioning matters remains paramount.

References

1. United Nations Convention on the Continental Shelf, Art. 2.1 and 4. The UK is a party to the Convention. The Continental Shelf Act 1964, s 1, confers on the Crown those rights of exploration and exploitation enjoyed under international law in respect of seabed minerals other than coal. The text of the Convention is given in A.D.G. Hill (ed.), *Daintith and Willoughby's United Kingdom Oil and Gas Law* (London, 1984), vol. 1, paras 2-003/031: henceforth this work is referred to as *United Kingdom Oil and Gas Law*.
2. United Nations Convention on the Continental Shelf, Art. 5.1 and 6.
3. For the answers of geographers and geologists see *United Kingdom Oil and Gas Law*, vol. 1, para. 1-104.
4. UNLOSC 1982, Art. 76.1. The outer limit of the continental margin is also defined in terms of distance: Art. 76.5. For the text of UNLOSC 1982, see (1982) 21 *International Legal Materials*, 1261.
5. [1985] I.C.J. Reports 13. UNLOSC 1982, Arts 55 and 57 delimit the outer limit of the EEZ at 200 nautical miles from the baseline of the territorial sea.
6. On the EEZ, see D.J. Attard, *The Exclusive Economic Zone in International Law* (1987); E.D. Brown, *Sea-Bed Energy and Mineral Resources and the Law of the Sea* (1984).
7. The Continental Shelf (Designation of Additional Areas) Order 1974, S.I. 1974 No. 1489.
8. The figures for installations are taken from *United Kingdom Oil and Gas Law*, vol. 1, para. 1-759.
9. Geneva Convention 1958, Art. 5.4; Continental Shelf Act 1964, s. 11 (prosecution of offences); Oil and Gas (Enterprise) Act 1982, ss 22 (application of criminal law) and 23 (application of civil law).
10. Continental Shelf Act 1964, s. 4(1) applying the Coast Protection Act 1949, s. 34.

11. This is somewhat circuitously achieved as an adjunct to the requirement that wells can only be abandoned with consent. See *United Kingdom Oil and Gas Law* vol. 1, para 1-776.
12. Under the Petroleum and Submarine Pipe-lines Act 1975, s.21.
13. *Essoview*, December 1995, 3. This need not be at the end of a field's life.
14. Henceforth referred to as the IMO *Guidelines and Standards.*
15. *Encyclopaedia of Scots Law*, vol. 18, para. 547; *Gloag & Henderson The Law of Scotland* (10th edn), para. 37.5.
16. In England reparation is for damage caused by tort and in Scotland where the harm is caused by a delict.
17. *United Kingdom Oil and Gas Law*, vol. 1, para. 1-778.
18. June 17, 1995, 1.
19. Department of Energy, *Brown Book* (1987).
20. See H.R. Dundas' "Abandonment of Offshore Installations: Current Legal Developments and Some Practical Considerations" (1992), para. 4.5.3: paper presented in the Centre for Petroleum and Mineral Law and Policy, University of Dundee.
21. Ordinarily, a private party injured through a state's breach of international law has no right of action against that state. For a more detailed analysis of state responsibility for decommissioned installations and for the defence of sovereign immunity, see *United Kingdom Oil and Gas Laws*, vol. 1, paras. 1-781/83.
22. *United Kingdom Oil and Gas Law*, vol. 1, para. 1-764.
23. *United Kingdom Oil and Gas Law*, vol. 1, para. 1-762. For discussion of the view that Art. 5.5. has fallen into desuetude (ie., disuse), see P.V. McDade "International Law of Abandonment of Offshore Installations: A Reassessment" (1985/6) *Oil and Gas Law and Taxation Review*, 291.
24. These are reproduced in *United Kingdom Oil and Gas Law*, vol. 2, paras 7-020/23.
25. Potential effect on the marine environment is determined on the basis of scientific evidence and, amongst other things, takes into account "the presence of endangered or threatened species": section 2.3.
26. Piper Alpha was left on site after consideration for the safety of those who would have had to remove it.
27. Critics of the plan to dispose of the Brent Spar were wrong to argue that deep-sea disposal in this instance set a dangerous precedent. This took no account of the Government's obligations.
28. Section 1.4. The UK has done so.
29. Section 3.13.
30. Section 3.6. This would allow a fully laden tanker to pass overhead without rupturing her hull. See H.R. Dundas, above n 20, para. 2.4.3. and fn. 21.

31. Section 3.4. (1).
32. *The Oslo Convention for the Prevention of Marine Pollution by Dumping from Ships and Aircraft* (1972). See *United Kingdom Oil and Gas Law*, vol 1. Para. 1-776.
33. Art. 5, Annex I.
34. Art. 6, Annex II.
35. *United Kingdom Oil and Gas Law*, vol. 1, para. 1-766.
36. Art. III. 1(a)(ii).
37. In 1990 a committee of "Legal Experts on Dumping" concluded that such disposal constituted dumping. The following year the legal experts seem to have retreated somewhat from this view.
38. This brings the north European states into line with the states parties to the Kuwait Convention (1990), which makes provision for the leaving of pipelines on the seabed. See H.R. Dundas, above n. 20, para. 3.3.
39. Annex III, Art. 5.
40. The 1985 Act applies to "marine structures", which is defined (s.24) to include platforms and exclude pipelines.
41. 1987 act, s. 2(1),(2),(5). A fifty per cent holding in one company by another is enough to bring both within the purview of the Act.
42. 1987 Act, s. 1(4).
43. 1987 Act, s. 4.
44. 1987 Act, s. 5(1).
45. 1987 Act, s. 9(1),(3).
46. 1987 Act, s. 6.
47. 1987 Act, s. 7.
48. 1987 Act, s. 10(4).
49. *United Kingdom Oil and Gas Law* paras. 1-774 and 3-258.
50. 1987 Act, s. 11(1).
51. Above, references 19 and 20.
52. 1987 Act, s. 8.
53. 1987 Act, s. 9(1).
54. 1987 Act, s. 10(1),(2). This power may be exercised both before and after approval of an abandonment programme.
55. 1987 Act, s. 2(3),(4).
56. 1987 Act, s. 5(2).
57. 1987 Act, s. 9(1),(2).

58. 1987 Act, s. 10(1),(2),(3).

59. 1987 Act, ss. 12 and 13.

60. Petroleum and Submarine Pipe-lines Act 1975, s. 25.

61. Model clause 19(7) of a production license so provides. Model clause 19(6) prescribes the conditions under which ownership passes to the Crown. These are (a) that the well has not been abandoned and (b) so that it remains in good order and fit for further working or (c) has been plugged and sealed in accordance with the Minister's directions.

62. International Convention on Salvage 1989 Act, Art. 1(a). This convention, otherwise known as the London Convention, is given force of law in the UK by the Merchant Shipping Act 1995, s. 224 and Sch. 11 Part I. In the UK, however, the Convention does not apply to salvage on inland waters where the vessels are only of inland navigation or where no vessel is involved: Sch. 11, Part II, para. 2.

63. Art. 1(a). It has been held that a gas float could not be salvaged: *Gas Float Whitton No 2* (1897) A.C. 337.

64. Art. 1(c).

65. Art. 1(a). Note also Art. 8(a).

66. Arts. 13(1),(b) and 14.

67. In the case of pipelines the owner may be the Secretary of State. See ref. 59 above.

68. *Decommissioning Oil and Gas Fields*, para. 42.

69. Petroleum Act 1987, ss. 21-24.

6
Political Aspects of Decommissioning

A G Jordan and L G Bennie

Introduction

Discussion of the political significance of the decommissioning of offshore structures means encountering a number of fundamental concerns of social science. One of these is the relationship between two central terms: the *public interest* and *private interest*. Another is the relevance of the *consultation model* of political decision-making. Yet another concerns the reasons for policy change. The Brent Spar events during April of 1995 - up to the date of the Greenpeace occupation - exhibit the search for consent among the affected interests that underpins much, particularly technically based, governmental decision making. However, ultimately, this case indicates that the search for agreement is not necessarily fruitful if there are fundamentally different objectives, values and priorities among those with an interest. Moreover the success of the consultation may have been undermined by the lack of breadth in its range.

'Public' and 'Private' Interest

Some political solutions are held to represent the *public interest* but it can be argued that business systematically has a privileged position in decision making that is at odds with this notion. Is there a conflict between business dominance in decision making and the public interest? The oil industry view would be that - at least in this case - what is good for the industry is also the best deal possible for the public. Why should a commercial decision by an oil company be considered 'political' and a public concern? Firstly, there is an inevitable impact by Government on so called private sector firms: indeed there is a respectable argument that Government is currently as involved in something like the water industry as much post as pre privatisation. Government affects industry indirectly by its handling of the macro economy, directly by its taxation regimes, by regulation of matters such as Health and Safety.

There is a broader argument that locates the decisions by companies in the political system. Roman law has the 'doctrine of the wild anima': an owner has responsibility if the beast escapes, even if the accident was not directly the fault of the owner. For that reason the oil industry is seen as having duties in so arranging its affairs as not to *let the wild beast escape* to inconvenience or harm others. In more modern jargon there are externalities to business decisions that are public rather than private considerations. Thus the *Economist*'s leader on Shell's decision not to sink the Brent Spar was, 'When you fill your car with petrol, or buy a pair of jeans, you are not taking a political decision. Or, at least you think you are not. But what if the firm that sells the petrol is a firm that dumps toxic waste in the ocean?' Business leaders take decisions in the economic interests of their companies rather than society in the broad, but as Meenan suggests in Chapter 2 is naive to assume that without specific incentives industry will 'be good boy scouts and break camp leaving nothing but ... thanks'. Perhaps the idea that industry will always take public spirited decisions without inducement to do so is as unacceptably sweeping as the notion that business always prevails in the political process.

The history of decommissioning in the North Sea is open to conflicting interpretations. On the one hand there is a view that Greenpeace successfully intervened on behalf of the public interest to 'sink' a policy self-interestedly aimed at minimising costs to the operators by deep sea dumping of redundant oil industry equipment in the Atlantic. This view sees the process by which Shell reached agreement with the Government as antithetical to democracy. However the alternative view is that Shell after extensive work satisfied the scrutiny of the public authorities of the merit of their proposed solution; and that this environmentally and democratically sound option was frustrated by the actions of an unelected and unaccountable Greenpeace. This line of interpretation sees the Brent Spar episode as a high point in a dangerous tide of pressure group power. The Greenpeace option, in this view, will lead to on shore recycling that will be more expensive, more dangerous, costly in terms of energy requirements and may harm coastal areas.

By 1996 (*Daily Telegraph*, Oct 12th) it appeared that at least some senior staff in Shell were surprisingly close to endorsing the former view. Cor Herkströter, chair of the Anglo Dutch group's committee of managing directors, said that Shell's 'technological arrogance' had left the group ill prepared to cope with 'growing social and political pressures'. He said that Shell had been slow to appreciate the growing authority and importance of environmental and consumer groups and 'failed to engage in serious dialogue' with them. He said that Shell had focused too heavily on internal matters and 'we failed to fully understand the need to provide information to the general public. ... we tended to become inward-looking, isolated and consequently some have seen us as a "state within a state"' [i]*. This reaction is significant as it concedes there is a problem. Shell initially, and many in the oil industry still, think the problem is better communicating the 'rightness' of the dumping decision: better PR rather than better policies.

The Greenpeace occupation of the Brent Spar took place on 30th April 1995. The political significance of the series of events from Greenpeace action to Shell's

---

*See notes at the end of the chapter

volte face has been much greater than suggested by the scale of costs involved [ii]. It has (frequently) been termed a 'defining moment' in relation to business and environment. Shell was one of the largest companies in Europe and in a poll in the *Economist* in 1995 it was rated the best-managed major oil company. If Greenpece could 'turn over' Shell, then the environmental cause was seen as assuming an enhanced importance. Many observers have concluded that on the strength of the Greenpeace intervention many businesses, and not just the oil companies, have increased the weighting given to environmental considerations in major decisions. The *Daily Telegraph* (22nd June, 1995) commented that the story, 'posed uncomfortable questions about the extent to which pressure groups can set the agenda'. It quoted a Government official asking, 'If Shell cannot outface Greenpeace, who can?' The chapter involves an assessment of the consequences for business and for Greenpeace, and discusses the wider implications for democracy.

The Consultation Model of Political Decision Making

Traditional accounts of British politics emphasise the formal institutions of Parliament, Cabinet and Prime Minister as the centres of political power. In fact there is another sort of politics that is played under rules that are quite different. In this sort of politics solutions are arrived at by negotiations between Whitehall and relevant interests. This is the politics of consultation. As Ian Greer, the most prominent British lobbyist put it [1, p. 110], 'many of the problems of clients are best resolved in discussions with civil servants before a government commits itself in public on a policy'. In this form of politics one aim is to *de emphasise* the *apparent* importance of what is going on. The 'rule' is to suggest that the policy is of limited interest and hence can be satisfactorily resolved by civil servant/group or firm discussion. In this sort of politics there is some merit to seeing all those bodies in attempting to determine policies as '*policy participants*'. To label every organisation attempting to exert influence a 'pressure group' causes confusion when in common currency that term usually means a multi-member collective body - unlike the business firm which is nonetheless often very active in trying to perform the pressure group *function* . Often the subject matter of consultation politics is fairly esoteric - though, as noted, it suits participants to present the issues in this low profile way and they may be of prime importance to those affected. It can be argued that in fact much of the substance of political activity is the aggregation of these rather esoteric sounding details.

A key phenomenon concept in political science is that of the *sub government or policy community.* This implies that decisions are not made in the large and general 'Government', but in parts of the organisation often in close contact with non Governmental organisations. There is an idea of loosely coupled systems. Most political decisions are *not* reached at cabinet level by politicians making partisan judgements, but by civil servants and interest groups finding agreement. This chapter broadly assumes that the decision process that lead to the licensing by the DTI of Shell to dispose of the Brent Spar in the ocean is an example of a subgovernment in practice.

Stewart [2] and Truman [3] identified a group's ability to supply policymakers with information as 'one important factor among the informal determinants of access' [3, p. 333]. Truman [3, p. 334] maintained that such information or knowledge can be divided into two main types - technical and political:

> *... technical knowledge that defines the content of the policy issue; and political knowledge of the relative strength of competing claims and the consequences of alternative decisions on a policy issue ... where official sources of information are deficient, command of technical knowledge may provide access for groups that can supply the deficiency.*

The policy making process operates thus because Departments require relevant *information* (which the groups possess), and *consent*. Civil servants prefer to have the groups 'on side', because of a general belief in the legitimation of policy by those affected, or more pragmatically as a method of pre-empting criticism and conflict avoidance. 'Down sizing' in Whitehall has only reinforced the situation that Departments often lack the intimate knowledge of sectors in which they wish to intervene; for that reason they have to try to 'de bug' proposals by using the expertise in the affected area. Pressures toward close group bureaucratic links show a 'logic of negotiation'. Given the demands on civil servants and group officials, this kind of policy making is mutually advantageous and there is the development of exchange relations.

There is also undoubtedly a cultural/constitutional convention that holds that policy making is more legitimate when affected interests are involved, and ideally satisfied. Civil servants also look to interest groups to help deliver political support. Thus consultation is a functional necessity in the process of developing effective policies. In the 1980s it also became increasingly the case that consultation was the norm as civil servants sought to avoid the scope of judicial review. That negotiations tend to be carried out in secret or at least on a confidential basis is as Eckstein maintained [4, p. 158] not due to any 'anti-democratic collusion', but merely because few people have an interest or knowledge of many of the specific technical issues being considered. Critics of the Government's granting of permission to 'dump' in the Atlantic often pursue the line expressed by the *New Scientist* (24 th June, 1995), 'Because ministers and oil companies opted for a cosy cabal, they have been hit by a Greenpeace hue and cry'. But in fact from the point of view of the Government and the industry the issue was not that they were in a cabal but no one else was interested. There was very little press or other attention to the February announcement. The international consultations that DTI were obliged to make attracted no response.

Extensive consultation is an integral part of the 'management of pressure'. Civil servants find that it takes less time and effort to agree to have discussions with a group than to refuse them consultation rights. This process may be purely cosmetic, but it allows civil servants to present an image of wide participation. Not all groups can be incorporated by these practices: some groups are saying things that are simply not politically acceptable. These (often) value changing proposals are not the stuff of policy community acceptance. If a group has a view that would make a substantial/controversial change to existing practices (e.g. animal welfare demands) they must expect to make their decision making input elsewhere (at least initially) in the political system. On any particular topic the consultation list used will reflect a judgement by the relevant civil servant. Civil servants seem to find little difficulty in sorting out the 'key groups'. *No one seriously proposes that all 'groups' are equally significant on all issues.*

One particular and well developed argument about the deficiency of the consultative system is that it is the sort of policy making machinery that entrenches the advantages of the well organised business forces. This argument about the privilege of business has been most famously set out in the work of Charles Lindblom [5]. But is business monolithic? Connelly has made the point that the multiplicity of business groups is a weakness rather than a strength [6]. He discusses the discussion paper on eco-labelling in 1989 and notes that there were 40 trade associations and 20 manufacturers and retailers on the list of 94 respondents to the paper. He argues that the apparent predominance of business was less an advantage than reflecting the fissiparous nature of the representation. In the Brent Spar case an important dimension was that not only was business not united, even Shell itself was fundamentally riven with the crucial power over Shell UK which forced a change in policy coming from the other parts of the Shell 'family'.

The 'privilege of business' interpretation sees business as attaining its goals because of the way in which it can manipulate the public agenda and instil generally pro-business views. Yet, in fact, Dymond's data about who the public trust on the environment shows environmental groups (including Greenpeace) are trusted by 81-85% of the public whereas the oil industry is trusted by only 32-36% and spokespeople for the oil industry by only 25%. A UKOOA briefing paper in November 1995 showed survey findings where the British public were more than twice as likely to rate the effectiveness of the oil industry (36%) at getting their viewpoint across than environmental groups (73%). In practice, the systematic pro-business indoctrination that is assumed by those who regard business as in a privileged position has not saved business from press and public cynicism.

From Policy Community to Issue Network: Unsuccessful Policy Resolution?

The essence of the consultation type of policy process is that 'out there' in society is the information and expertise which Government needs to make sensible policy. Moreover a policy which is the product of a process of looking for the cooperation and consent of affected interests is thought to be more legitimate than one foisted on affected interests. This chapter assumes that the consultation process is a normal style of policy making and the Brent Spar story can be fitted into the description quite comfortably - until 1995. Whereas for the British public decommissioning was 'invented' as an issue by Greenpeace in 1995, in fact there had been various conferences, consultation schemes and meetings in the industry in the previous decade.

But ultimately this episode shows the limitation of the usefulness of policy community style politics. The 'insider' agreement was overturned by Greenpeace action. Critics would see the fact that Government initially accepted the view of Shell was not so much an example of the Government finding reliable advice, as the industry 'capturing' the Government. If consultation is successful where it produces viable agreement, then this example was a failure as there was no robust consensus. The events in 1995 more nearly accorded with what has been termed an *issue network* interpretation of politics [7]. A large number of groups were mobilised; there was no trust or patterns of stability that would allow trade-offs over time among participants; there was conflict over underlying values and a lack of a shared appreciation of issues.

Ultimately successful or not, consultation was as big a feature of policy development here as in most other policy areas. The DTI was the 'lead' Department in Whitehall as the Government's evidence to the House of Lords Committee in 1995 shows (HL Paper, 114, 1996). The DTI chairs the inter-Departmental Decommissioning Policy Review Committee which includes representatives from the Ministry of Agriculture, Fisheries and Food, the Ministry of Defence, the Department of the Environment and the Scottish Office Agriculture, Environment and Fisheries Department, and a range of other departments with more peripheral interests. The Brent Spar abandonment scheme was 'processed' in Whitehall in the usual procedures.

When North Sea structures are deemed non-viable, the Government reviews - and must approve - the disposal plan. Guidance notes have been developed by Government which say that a 'framework for orderly abandonment' is needed. Partly this is to avoid damage to the interests of other parties in the North Sea (most particularly the fishing industry) and to avoid the Treasury having to cover a bill some time in the future. Consultations are required and this should include, 'those interested parties ... who may be affected by the abandonment programme'. The DTI Consultative Document on the Abandonment of Offshore Installations and Pipelines under the Petroleum Act 1987 (May 1995) suggests an illustrative list of bodies the DTI may require operators to consult. In Scottish waters this is: Scottish Fishermen's Federation; Orkney; Fishermen's Association; Western Isles Fishermen's Association; Firth of Forth; Fishermen's Association; Scottish Natural Heritage; Seabirds at Sea; British Telecom International.

Operators must provide a 'Best Practicable Environmental Option' as part of their decommissioning case (or more precisely, a BPEO has been taken to be the sort of scheme that would give Government confidence in the proposal). A BPEO has been defined by the Royal Commission on Environmental Pollution as 'the option that provides the most benefit or least damage to the environment as a whole, at an acceptable cost, in the long term as well as the short term'. The BPEO became the standard against which the UK Government will judge all applications for licences for the decommissioning of all oil and gas installations. The Royal Commission also requires appropriate and timely consultation with people and organisations directly affected to be taken into account during the BPEO procedure. The BPEO was based on a study conducted by AURIS in January 1994, 'The Removal and Disposal of the Brent Spar' and in December 'Environmental Impacts of two Possible Disposal Options for the Brent Spar' and 'A Risk Analysis of Two Possible Disposal Options for the Brent Spar Buoy'. These reduced 13 potential abandonment options to two that warranted further study before concluding that deep sea disposal was preferable. In this view the Shell case was evidence based and not simply a cost containing solution.

The Shell account of the processes they followed in reaching the BPEO is set out in the report of the Scientific Group on Decommissioning Offshore Structures (22nd May, 1996). One feature that comes from that account is that Shell worked closely with Governmental scientists in selecting topics to investigate and in the analysis of results. Shell said, ' ... when the survey had been completed the Scottish Office scientists conducted a rigorous analysis of the data, and concluded that the environmental consequences had been adequately assessed for the three proposed sites. After reviewing the data, *the Scottish Office scientists selected the disposal site ...*' (emphasis added, Shell NERC document, July, 1996). (The approval of the abandonment plan is a DTI responsibility but the intended location meant that the Scottish Office granted the deep sea disposal licence.)

The commonplace interpretation of the Brent Spar story probably assumes that the response of Greenpeace 'came out of the blue' in the form of direct action on the Brent Spar; that as Shell was going about its legitimate business an environmental group disrupted their plans. In fact, Greenpeace had been busy lobbying for a number of years. For example, in November 1993 Greenpeace was active [iii] at the London Convention [iv] Amendment Group which dealt with matters such as the disposal at sea of radioactive waste and industrial waste. Significantly even then they cited the 'Precautionary Principle' agreed at the 1992 Earth Summit which encouraged Governments to, 'apply preventive, precautionary and anticipatory approaches so as to avoid degradation of the marine environment, as well as to reduce the risk of long-term or irreversible adverse effects upon it' (Agenda 21, 17.22(a)). The Precautionary Principle had also been endorsed at the London Convention itself in 1991 (LDC.44 (14)).

On February 16th 1995 Shell announced the DTI approval for deep sea disposal. Greenpeace's response was to release a review of the BPEO by Ruddall Blanchard

Associates Ltd for Shell Expro. This apparently concluded that the AURIS survey did not necessarily mean that the conclusions should be treated as definitive. Though Shell obviously thought their research was adequate, Greenpeace condemned the DTI licensing on the basis of 'meagre information' (*New Scientist*, 23/03, 1995). Greenpeace was able to cite international obligations that put the deep sea disposal in question. The 1958 Geneva Convention on the Continental Shelf covers the North Sea. This sets out that, '... any installations which are abandoned or disused must be entirely removed' (Article 5(5)). This was only one of the national and international guidelines and regulations that apply. Others include the Oslo [v] and London Conventions Guidelines (1991), Oslo Convention (Convention on the Prevention of Marine pollution by Dumping from Ships), the International Maritime Organisation's (IMO) Guidelines and Standards for the Removal of Offshore Installations and Structures on the Continental Shelf and the Exclusive Economic Zone (1989), UN Convention on the Law of the Seas (1982, Article 60(3)), Convention on the Prevention of Marine Pollution by Dumping of Wastes and Other Matters - London Dumping Convention (1972).

The status of all the international agreements is not entirely clear, but Greenpeace was almost certainly correct in their argument that it was at least 'pleadable' that dumping was prohibited - or at least against the spirit of a developing tendency. A report by the Parliamentary Office of Science and Technology argued (July, 1995) that a trend against marine pollution 'has been strengthened by agreements at the North Sea Conferences and has lead to industrial waste dumping being phased out altogether this year, and an agreement to phase out sewage sludge by 1998'. As Chris Rose of Greenpeace said in a letter to *The Times* (24th June, 1995) if the Brent Spar waste was removed and packed into barrels, international obligations would have meant that it would not have been acceptable to dump it at sea.

In March 1995 Greenpeace (19th March) recorded that disposal of the Brent Spar was expected to take place just prior to a meeting of European environment Ministers in Denmark on the 8-9th June. Some critics of Shell say that it was predictable that Greenpeace would have wanted to make some public point before the conference and Shell should have been prepared. At this time the Labour Party came out against deep water disposal on the grounds that the decision represented 'short term economic benefits'.

On 30th April Greenpeace occupied what they termed 'an old North Sea oil platform to stop its owner, Shell, from dumping the rusting hulk and its highly toxic contents to the sea bed' (Press Release). The Brent Spar was deserted and without security though a press release dated 29th April reported that their boat the 'Moby Dick' was heading for the 'dilapidated Brent Spar oil rig ... This time, the target was the oil companies polluting the North Sea'. The Greenpeace case was based on 6 main arguments:

1. There had been no proper inventory of the Brent Spar contents, so the environmental impact could not be assessed.

2. The impact on the deep sea environment was unknown.

3. The documentation on which Shell's application to the UK Government was based was inadequate.

4. Dumping would be a precedent.

5. Dismantling was technically feasible and 'quick and dirty dumping', if permitted, would reduce the incentive for innovation in recycling and reuse.

6. In principle recycling should be followed. (see Press release 20th July 1995).

The first Greenpeace press release claimed that the Brent Spar contained over 100 tonnes of 'toxic sludge' (16th June, 1995). The press release not too subtly tried to discredit the Shell case by pointing out that, 'The UK Government's decision to allow Shell to dump the Brent Spar was based on information supplied and paid for by Shell' [vi]. Media treatment of the occupation was characterised by one example cited by Dr Chris Fay of Shell, 'Murder at sea - scandal of dumped rigs that may wipe out marine life'. The structure was occupied for 23 days until Shell security (with police observers) boarded and removed the protesters (military assistance was available but not required).

During and after the occupation Greenpeace kept up a PR offensive with a series of claims that captured the headlines and made sure that the media event was reported in ways that discussed a Greenpeace agenda. For example they released a report that said the on shore disposal would not be as expensive as Shell claimed: they famously asserted that there were over 5,500 tonnes of oil aboard. Shell was usually left with a reactive damage limitation response. The claim was subject to a brief Shell release, '(Shell) believes that this is yet another example of how Greenpeace are wilfully misleading the public with false and unsubstantiated information' (June 17th). Greenpeace was constantly changing the nature of the argument by new claims. Shell's attempts to counter with detailed arguments did not alter public perception of what was going on in the North Sea. As late as a public opinion poll for Greenpeace in February 1996 it was found that 74% of the public backed a continued Greenpeace campaign against the dumping of oil platforms. In the British collective memory there seems to be a view that Greenpeace secured heavy and very favourable press coverage during the occupation and that the Greenpeace claim about the 'concealed' oil was influential. In fact the press attention during the actual occupation was remarkably slight. The press treatment built up after the occupation and picked up on developments in Europe. The claim about the oil content was not made until the 16th June. This was hardly crucial as Shell was already accelerating down the slope to policy change. So the 5,500 tonnes of oil claim did Greenpeace little good - and eventually soured their relations with the press to a remarkable degree.

Having regained control of the Brent Spar, Shell conducted work to prepare for final disposal. By 14th June the 'reclaimed' Brent Spar was under tow to its North Atlantic disposal site, but while the Head of Shell in Germany, Peter Duncan, had initially supported the 'dumping' option, Greenpeace was able to report that protest had rapidly expanded in Germany. The position of Shell Deutsche moved in response. Duncan became an in-house critic of continuing the deep sea disposal. Polls for Greenpeace Germany said that 74% of the public and 85% of car owners would boycott Shell (Greenpeace, 14th June, 1995). This public reaction was exploited by Greenpeace who made a call for motorists in Britain to join in the boycott (15th June). Reports of a 50% (elsewhere reported as 30%) down turn in sales in Germany of fuel were made and Chancellor Kohl raised the issue with the UK Government.

Shell in the UK put out a press release on June 17th describing how the Greenpeace day of action at its service stations 'barely occurred'. Shell's complacency in Britain was exposed when pressure on the sister companies meant that they were compelled to change their plan hours before the disposal was due to take place. Shell's release missed the important fact that Shell was vulnerable to behaviour outside Britain. The *Boycott Quarterly* (Fall, 1995) recorded that, 'The boycott of Shell was, without question, one for the record books, and the people of Europe taught one of the largest companies on Earth a ten day lesson in the economic theory of supply and demand that may have changed the battlefields of the activist community forever'.

On June 20th Greenpeace was 'elated that Shell reversed its plan to dump the "toxic-laden" oil installation'. It reported Shell's announcement that it had abandoned its deep water disposal plans and asked the UK authorities for a licence to dispose on shore. This switch put the British Government in an embarrassing position as they had publicly stuck with Shell through the protest, and indeed in Prime Minister's Question Time that day John Major had defended the (abandoned) Shell position. The Shell Press Release on 20th June expressed the belief that Shell UK Ltd '*still believes that deep water disposal of the Brent Spar is the best practicable environmental option, which was supported by independent studies*'. Soon the Shell tone was to change as they inched towards saying that the deep sea option was good, but something even better is probably possible. In the June 20th climb down Shell defensively reiterated that it had Government backing for deep sea sinking and that their conduct had been in accordance with established international policies and standards, but they noted that other governments, 'have taken an interest and voiced strong objections' and, 'Notwithstanding the efforts to convince these governments of the validity of the approach, most remain strongly against deep water disposal'. The release then went on to imply that it was the attitude of governments that had made difficulties for Shell operating companies, 'The European companies find themselves in an untenable position and feel that it is not possible to continue without wider support from the governments participating in the Oslo - Paris Convention'.

Undoubtedly there was bitterness by Conservative Ministers that they had taken political attacks on behalf of Shell, only to be embarrassed when Shell ditched the policy and left Ministers defending an abandoned policy. Ministers also seem to have been bitter about the lack of support from other European countries. Hints emerge frequently that the reaction would have been different if France, Germany or Holland had had significant off shore oil resources. At the meeting of Ministers in June most of the North Sea states accepted the banning of dumping, but the states primarily affected (UK and Norway, with France) opposed. This allowed Greenpeace to attack the UK Government and Shell as continuing to 'view the North Sea as their private dumping ground' [vii]. The Ministerial Conference agreed (with reservations by Britain) to 'continuously reduce discharges, emissions and losses of hazardous substances thereby moving towards the target of their cessation within one generation (25 years)' (Greenpeace, 26th June, 1996).

On July 7th Shell was given formal approval by the Norwegian Government to moor the Brent Spar temporarily at Erfjord. In July the same month Shell invited DNV to conduct an independent inventory on the Brent Spar. This reported in October [viii] - confirming the broad Shell position. In early September (5th) Greenpeace publicly admitted that its claim about the scale of oil sludge was wrong. The emotive oversimplification by the press that had greeted Greenpeace's claim was now to Greenpeace's disadvantage. The story was treated in terms of 'Greenpeace woz Wrong'. In fact Greenpeace could legitimately claim that they were given misleading information and more importantly they could claim that the essential case did not rest on the volume of material to be 'lost', but the damage was certainly done to their credibility and they sounded unreliable and sloppy. However in determining the policy change more important than Greenpeace's lie/mistake were the claims that they made based on Shell's own data - that there were over 100 tonnes of residual chemical sludge on board. This was Shell's own data but Greenpeace presented this as a toxic threat whereas most the chemicals woud not have been out of place on a seabed. It was the 'spin' that they placed on the accurate figures that was perhaps more influential than the inaccurate.

Reinventing a Policy Community?

As against the vehement anti-Greenpeace noises made by Shell in mid 1995 when they thought they were the victims of unfair Greenpeace tactics and claims [ix], a remarkable conversion has taken place as Shell clearly has concluded that, like it or not, they must 'do business' with their critics. Greenpeace met Shell on 6th September 1995 and welcomed a 'constructive meeting' between Dr Chris Fay, Chairman of Shell UK and Peter Melchett Executive Director of Greenpeace. A further meeting between Thilo Bode of Greenpeace International and Peter Melchett of Greenpeace UK and Shell took place on 9th January, 1996.

One reason for some emergence of common ground between Shell and Greenpeace was the fact that Shell still has to gain the approval of the UK Government for relicensing of any on shore dismantling and disposal. The new review (of more than 200 options) had to identify the Best Practicable Environmental Option. The new proposal would have to match or be better than the deep sea disposal which, 'balanced safety and the risks to human life, technical feasibility, economics, and environmental impacts'. Shell agreed that in the consultations that would be part of developing a new BPEO proposal Greenpeace would be one of the representative bodies involved. On October 11th 1995 Greenpeace welcomed Shell's announcement of an open consultation process' (the Way Forward) and said it would be happy to participate in this consultation process. The 'sweetheart' noises between Shell and Greenpeace meant that Shell was adopting a very different tone from that of the Government, and in particular Tim Eggar. He presented events as an attack on science in which Greenpeace emerged with a credibility problem. He said, 'Blackmail succeeded against well proven scientific effort' (*Battle for Brent Spar*, 1995).

This is not to say that relations suddenly became warm with agreement on all fronts. Greenpeace has continued to seek to exploit their strong hand. In late October (24th) Greenpeace released a report by an independent consultant, Peter Brindley, that claimed that a coordinated on shore dismantling programme would be cheaper (by over one billion pounds) than the on-going case by case onshore/offshore pattern of decisions. Greenpeace submitted the report to the House of Lords sub committee on Trade and Industry (HL Paper 114) and the DTI's review of new guidelines on decommissioning. As late as October 1996 Greenpeace was claiming in its advertising material that financial support would help prevent 'an ocean devoid of life ... slowly filled with rusting toxic hulks'.

In November 1995 Shell was caught up in the controversy about the death sentence on Ken Saro Wiwa and eight other Ogoni in Nigeria. Greenpeace attacked the fact that Shell had 'flared' surplus gas in the Niger Delta for 37 years - contributing, they said, to lung diseases, still birth and polluted drinking water. Clearly the Nigerian embarrassment had nothing *directly* to do with the decommissioning story, but it interweaved the political campaigns of Shell and Greenpeace and the furore meant that Shell was more sensitive to the need to engage in some kind of public relations counter offensive. The Nigerian story undermined the Shell case that it always worked to the highest possible levels of environmental stringency.

The report of the Scientific Group on Decommissioning Structures (NERC) was published on 22nd May 1996 [x]. This concluded that the environmental impacts of deep sea disposal would have been small and localised and concluded that the idea that these impacts should have been avoided because of the 'Precautionary

Principle' was 'extremely tenuous'. Shell accordingly claimed that it broadly endorsed their own Impact Hypothesis (Shell NERC Document July, 1996). However Robin Grove-White, former Head of CPRE, and a Greenpeace board member claimed that an, 'NERC report had shown Greenpeace to have quite as good a case in the wider scientific frame as Shell and DTI had had within their narrower frame' (ENDS report 256, pp. 4-6).

Shell adopted the position that decisions adopted without some kind of public endorsement were likely to lack robustness and were therefore suspect. Accordingly they, and the industry through UKOOA, have engaged in a policy of 'open books' in terms of the new BPEO process. In line with this in 1996 Shell Expro 'launched the Brent Spar into cyberspace with its own Web site on the Internet'. The decommissioning manager, Eric Faulds, said, 'It should also help to ensure the ultimate decision and the reasoning behind it are widely discussed and understood'.

29 ideas from 19 contractors [xi] and consortiums for the Brent Spar were being considered by Shell 1996 (*Guardian*, 15th, 1996) - selected from 450 submitted to it. Only one was for dumping (in a specially dredged trench, a proposal from a Belgian dredging company). Greenpeace stated that they could support 21 of the 29. They objected to the dredging option on principle, but that they did not have adequate information on another 7. At a Greenpeace Business conference in 1996 (25th September) Greenpeace confirmed that they could live with most of the Shell options, but ironically this event was closed to the general media, TV crews or photographers. So the image of a 'clean' Shell received far less attention than the Brent Spar.

The First Seminar for Brent Spar Dialogue took place on 1st November 1996. Invitations from a 'neutral' body, the Environment Council, went out to interested parties (professional bodies, university researchers, consumer and environmental groups). The focus was on discussion of the 30 Long List proposals. Subsequently Shell , in early 1997, announced a short list of options and six companies were invited to submit costed options. DNV has been invited to comment on the options.

The issue of decommissioning is still being discussed by European governments. On 26th June 1995 there was a meeting of the Oslo and Paris (OSPAR) commissions in Brussels. A number of governments accepted a ban on dumping but the move required 10 of the 14 countries participating before it became policy. In the event UK and Norway were against a ban on ocean dumping. The UK objection meant that it was not legally binding on the UK. As the recommendation was non-unanimous an OSPAR working group (Sea Based Activities - SEBA) is meeting regularly with a view to Ministerial agreement in June 1997. The European Commission is waiting to see if OSPAR can reach an agreed conclusion before it decides if it needs to establish a 'competence' in this area. If there is consensus the Commission will leave this policy within the auspices of OSPAR.

Consequences of Brent Spar for the Oil Industry

Political success is often a matter of who succeeds in defining the situation in the public mind in terms that suit one side of the argument rather than another. Ironically in this sort of battle the apparently weak may be at an advantage: the small boat of the altruistic versus the large corporation was an irresistible media story that slanted things in favour of Greenpeace. When Shell used their resources to remove Greenpeace activists the imagery seemed to confirm exactly the Greenpeace interpretation: over mighty multinational trampling on the little people. The Greenpeace success was however not created by Greenpeace tactics and PR, but was because the Greenpeace charges resonated in sympathy with a public scepticism of business motivations.

Shell in this instance has taken the view that it is not willing to pay the image 'price' for following through with the dumping solution for Brent Spar that it had managed to successfully 'sell' to the Government. But more generally the industry wants to keep available the possibility of deep sea disposal for other structures. Accordingly there is a stress on managing public opinion to counter the Greenpeace interpretation that by definition 'dumping' is environmentally dangerous and wasteful. UKOOA have used the term 'extending public confidence'. The industry as a whole (through UKOOA and the Oil Industry International Exploration and Production Forum) have decided that they need to 'review the way in which it had approached decommissioning in the past. The industry recognises that there is a need for greater openness and transparency'. They expressed a new desire for 'dialogue with, and a new commitment to listen to, interested parties'. The processes set up by Shell are probably unprecedented in their thoroughness and breadth.

Shell have apparently moved some distance towards accepting the Greenpeace criticism. Dr Chris Fay conceded,

> *Bodies like Greenpeace clearly have the right to argue that offshore disposal is wrong in principle. We should remember the debt we owe to environmentalists for awakening society to the environmental challenges we face. They remain a vital strand of opinion in our environmental considerations.*

Of course if it is decided that the public consequences of not dumping the Brent Spar are not to be borne Shell needs the assistance of environmental bodies (especially Greenpeace) to help persuade the Government to approve a new plan. Having persuaded Government of the credentials of the dumping plan, Shell has to argue that they have an option that was better than the best. This has involved Shell in some deft foot work in now arguing that what were once inescapable constraints are now (resolvable) technical issues. By 1997 there may be a final twist to the saga. Shell may discover that the factors that pushed it towards dumping are so important that they make impossible the on shore solution that they now want for political reasons.

At a media conference Dr Fay of Shell said that the news coverage of the Brent Spar had featured independent oceanographers who tended to support the Shell case, but he acknowledged that, 'I don't claim that proves we were right. Some scientists tended to Greenpeace's point of view and an argument along the lines of "my scientist is better than your" scientist clarifies little'. In so doing, Fay was recognising that scientific 'facts' are rarely unambiguous. He noted that there was need for on-going activity to help keep the public in touch with the argument about the importance of industry in wealth creation. Whatever the good intentions here there has to be some doubt about this as a feasible option. The idea of securing the 'consent of customers' in the sense that consumers have a 'strike weapon' of non-purchasing is sensible. What can be done is have meetings with interested groups such as the meetings with Greenpeace under the auspices of the Oil and Gas Consortium and the Wildlife Link. But are these groups identical with the 'public'? To what extent does a policy of open discussion with such bodies really mean granting veto powers to these bodies? This is all well and good when a shared conclusion can be reached, but what weight is given to views that disagree with industry? Is agreement with such bodies the same as public approval?

The Brent Spar incident has been seen as transforming business attitudes to the environment: these issues are seen as too expensive to ignore, if not too important. If Europe's largest company was frustrated by an environmental group, then the signal seemed to be that all companies were vulnerable. So the significance of Brent Spar was not just that it was a reasonably big bit of engineering, or even that it was a precedent for several hundred other oil decisions, but that it was the pivot on which a more general business reappraisal of the environment took place.

Consequences of Brent Spar for Greenpeace

Why did the Greenpeace campaign prove so successful? Central to the strategy of the environmentalists was the use of the media to mobilise public support. The image of underdog standing up to the vested interests of industry was cultivated by Greenpeace and projected by the media. The media were more than willing participants in this partnership. A number of observers have noted the positive media image enjoyed by Greenpeace across Europe, and the importance of this relationship to the organisation [8; 9]. Rucht [8, p. 71] argues that gaining the support of the media is *the* primary objective of Greenpeace: all other functions are subordinate.

There can be no doubt that Greenpeace regard scientific/environmental knowledge as an important 'organisational weapon' (Eyerman and Jamison, 1989). In the

1990s Greenpeace UK have put more and more emphasis on the importance of scientific evidence to back up their campaigns. They have a scientific director, reports are distributed to scientific experts before publication, and press releases include references to scientific journals. This has been referred to as the 'new model Greenpeace' which has undergone 'a conversion towards science and the more conventional rigours of research' [9, p. 36]. Thus, the commissioning of scientific research has become a 'campaigning tool' for Greenpeace as much as their opponents.

The media respect enjoyed by Greenpeace is related to the organisation's use of scientific expertise to substantiate their more emotive arguments. Greenpeace presented their own technical scientific evidence to support their case. Some of those active in the oil industry might still feel that 'their' science was good enough to satisfy the Governmental requirements and that imprimatur should suffice. This neglects three points: One, there is rarely a black and white scientific answer. Secondly there is no level playing field. Because of past matters the public are more likely to find 'neutral' views of the environmentalists persuasive than 'biased' industry sources. Thirdly, if the Government is serious about claiming a 'green halo' for international commitments and the 'Precautionary Principle' then there seems an element of special pleading: these things apply - but not to us, at least not here and not just now. That is to say that Greenpeace's Precautionary argument is a good one - better than the toxic timebomb sort of scare story that perhaps underpinned their impact. When the Brent Spar activists argued 'Don't use the ocean as a dumping ground', Greenpeace's use of the term 'dumping' had useful PR benefits. As Forte suggests the more technical term 'abandonment' is only marginally better in its mental associations: it carries with it ('mother abandons child') the suspicion of failing to live up to responsibilities. Attaching the label 'dumping' to the deep sea disposal was as effective as stigmatising the Community Charge as the Poll Tax.

Greenpeace exerted considerable political influence in the campaign against deep sea 'dumping'. However, there was also the importance of pressure from international governments admittedly inspired by the environmental movement. And it could be argued that the real 'muscle' was not the skill or resources of Greenpeace and its financial supporters but the fact that the message was accepted by (European) consumers with no relation to the group. Ultimately it was *consumer behaviour* that worried Shell because that in turn put pressure on national governments that then acted on Shell national companies. The importance of Greenpeace was that it 'sold' an interpretation of events that lead to a consumer boycott. Over Brent Spar there was a message that concerned the public and the public could do something; not by running risks in cowboy and Indian games in boats, but in simply driving past the Shell forecourt to the next filling station. It was the arm chair environmentalists that beat Shell: the wet suited activists (mainly paid professionals) were the trigger not the explanation.

Despite the apparent success of the environmentalists' campaign, a number of criticisms have been directed at Greenpeace. For example, they have been associated with the violent attacks on 50 or so German Shell service stations. However, there is no evidence that Greenpeace was anything but distressed by an activity that was antithetical to the Greenpeace philosophy - and so spectacularly destroyed the carefully cultivated brand image of the 'good guy'. More serious questions were raised about the validity of the Greenpeace message. Their campaign has been described as 'misleading', based on 'misinformation', and they have admitted that some of their calculations were wrong. Matt Ridley of the *Sunday Telegraph* claimed,

> *Brent Spar was simply a high profile stunt pursued to reverse the organisation's declining membership. It raises money from its members and donors with the intention of spending £90m a year on things that will attract more members and donors while paying the salaries of its employees. (quoted in* Scotland on Sunday, *September 8th).*

In this cynical view the campaign was perhaps more about maintenance of Greenpeace as an organisation (see [10]). Income is linked to the public thinking that it is getting something for its money. The problem with quiet influence is in persuading financial supporters that they are making an impact. Even an unsuccessful stunt has more impact than successful insider negotiation.

In the wake of Brent Spar there was widespread belief in business that this was the face of the future and that business was to be seriously constrained. A book produced by Deutsche Shell (DSAG) (*The Events Surrounding Brent Spar in Germany*) saw Brent Spar as, 'a prototype for future Greenpeace campaigns and could in some aspects become an example for other environmental organisations to follow'. However, the most significant feature of Greenpeace's life post Brent Spar has not been its high profile activism but a new style of persuasion and business co-operation. The new Executive Director, Thilo Bode has taken the group in the direction of 'respectability'. He has talked of the need for 'real professionals. We should be at the forefront of solutions and technological innovation'. This has lead to portrayals of the tension within Greenpeace in terms of the suits versus the radicals. One internal critic has complained that direct action has become a marketing tool rather than a political tool and suggested there is a danger in just becoming a stunt-making, money-making machine (*Telegraph Magazine*, 1996).

Shell has adopted a good neighbour policy to Greenpeace, but more surprising has been the tone of some of the Greenpeace comment. The Brent Spar was not the start of something by Greenpeace but perhaps the tail end. Though now in regular contact with the oil industry, Greenpeace still attempts to create political capital (and income) from the Brent Spar affair: one recent leaflet asks the prospective Greenpeace supporter, 'Who organised the campaign - and won the moral victory - to stop the Brent Spar from being dumped?' (October 1996).

Consequences of Brent Spar for Democracy

The decommissioning policy case study raises a number of questions about political power and democratic accountability in Britain. Central to the debate is the role played by government. The image of what has been happening in 1996 is one of direct competition between industry and environmentalists for the support of the media and public. The government, meanwhile, has chosen to watch developments unfold from the political sidelines, unwilling to endorse the Shell policy switch and working to keep deep sea disposal as an option in discussions at international level. Chris Rose of Greenpeace has talked of 'politics without politicians - driven by consumers, technologies, business people, major retailers. The Government didn't control Shell, the customers did. Authority was exposed, as it was shown that the conventional political process was not running the country' (*Telegraph Magazine*, 1996).

The consultation stage in 1992-5 seemed to suggest that it was the business groups who, after all, have most influence. Those who argue for the use of consultation processes rest a large part of their case on the fact that they allow the participation of outside interests and legitimate the policy making process. Peters and Barker [11, p. 1] have commented that whereas once the concept of democracy implied primarily having elections at appointed times and allowing winners of those elections to govern until the next election, democracy has now acquired a more continuous character; and receiving advice helps governments to appear more open and democratic. So the core argument seems to be that a requirement of both an efficient and a democratic decision making system is that there be direct involvement and agreement of those who are principally affected. The problem was not that the Brent Spar decision initially came out of a consultative process, but that the machinery was not well handled. Partly because this was a fairly esoteric subject with few critical bodies prepared to invest resources in the topic, but partly because the matters was seen as requiring commercial confidence, the abandonment proposal was effectively taken in what Baumgartner and Jones call a 'policy monopoly' [12, p. 6]. They define this as when, 'Experts in all areas spend much of their time convincing others that "outsiders" are not qualified to make decisions in a given area'. This meant that the proposal was not subject to the cross examination by other interested parties that is the main quality control mechanism in full blown consultation. Even those who saw a strong environmental case for dumping identified at least three weaknesses in the Shell argument. These defects may not have been of such a severity that the case for dumping was undermined, but it meant there was a scientific 'noise' that Greenpeace could exploit (see Dr Tony Rice in *Guardian*, 22nd July, 1995). Furthermore the Government in adjudicating the merits of Shell's proposal had a financial stake in cheapness that was even greater than Shell's. Through tax relief the Treasury was principal paymaster of the operation.

If Government was 'in the pocket' of business the mobilisation of environmental interest groups might be seen as enhancing democracy. Environmentalists can be seen as a counter-balancing force to business interests. Greenpeace made precisely this point while justifying their direct action. In an interview in *Frankfurter Allgemeine* on August 18th, 1995, Thilo Bode said,

> *Democratic politicians are no longer able to make important decisions. The Western democracies are no longer able to manage the status quo; in a conflict of interests, they are nowadays only able to manage the status quo... Show a politician who is prepared to say that he thinks the incessant growth of industry is damaging ... With our controlled violations of legislation, we promote the development of rights. Democracy honours this method, it retreats, it makes the laws that we think necessary for the protection of the environment. Our law-breaking is the only language which is understood - it is essential for legislative development.*

However, there is also a potential danger of unrepresentative, unelected and unaccountable environmental groups exerting too much influence. It is clear that in future corporate and governmental decisions there will be allowance for anticipated reactions by Greenpeace. The irony is that these consequences of environmental action will never be visible.

The Brent Spar episode raises issues about the role of the media. Graber [13, p. 99] suggests that environmental claims are judged less critically than those of business and that because 'getting information quickly to the public' is so important some stories are not subject to any critical interpretation at all. Inevitably there is a bias in that 'bad news' treatments are inherently more newsworthy than 'Man does not bite dog'. Deutsche Shell (1995) complained of a journalistic sloppiness - or even a journalistic competition to give the public the kind of anti-business story that the public like. It said, 'The decline of journalistic accuracy is noteworthy. "Drilling platform" and "North sea" sound better and more suggestive than "storage and loading platform" and "Atlantic"'. It claimed that a photo of the work platform next to the Brent Spar was deliberately miscaptioned as the Brent Spar. *The Times* concluded in 1995 (10th September) that Greenpeace was 'smarter, better organised, and more popular than one of Europe's largest and most respected companies'. However our interpretation warns against giving Greenpeace more credit (or blame) than it deserves. At one stage during the occupation Thilo Bode complained 'The whole Brent Spar thing really isn't working'. In a sense Greenpeace failed according to the game plan of securing change by pressure on the British Government. Neither the Government nor British public opinion secured change. It was the unforseen sensitivity of the European consumer that was crucial. The misidentification (accidental or otherwise) by Greenpeace of the dump site as in the North Sea rather than the Atlantic aggravated these fears.

There were second opinions in the media about the uncritical treatment of the convenient Greenpeace version of events. Not only did Greenpeace spoon feed the press with press releases on more or less a daily basis, they also offered 'video footage and stills'. The head of BBC television and radio news commented, 'In some senses we were "had" over Brent Spar. I'm left feeling Greenpeace was pulling us by the nose through too much of the campaign' (see *Scotland on Sunday*, September 8th, 1996).

Conclusion

In the final analysis the decommissioning story fits neither the traditional elections/party politics model of decision making nor the consultation model. The process at work was a more 'open', messy type of politic. The sort of politics looked at under the heading of consultation is often termed a *policy community* approach: the relevant participants are restricted in number and tend to share a broad understanding of the issues in the sub sector. As noted the decommissioning example has many of the features of what Heclo described as an *issue network* [7]. As in the policy community arrangement there is no strong party political element, but the politics in issue networks are open, messy, conflictual: it is a battle not about the detail of policy, but about the sorts of values that should be considered as being important. A policy community interpretation of policy making seems to account for stability but does not well deal with instances of change such as over the Brent Spar. To a remarkable extent the episode validates the Baumgartner and Jones argument about *venue shopping* [12, p. 36]. Greenpeace succeeded in transferring the decision from the British policy community venue in which it had lost to an issue network involving an international multitude of conflicting participants. This literature on a competition by interested parties to determine the public interpretation of events is an important 'add on' to policy community writings. The issue network portrait on the post occupation stage captures a sense of chaos and conflict. This however was not the spirit that underpinned relations by the autumn of 1996 when the regular links between Shell and the environmentalists say something of the shared understanding of the feasible that was reminiscent of the policy community idea. This is not an unusual phenomenon. As conflict is resolved the process often borrows some of the characteristics of the policy community.

The chapter assumes that *direct relationships* between industry and environmentalists are based on serious intent at mutual accommodation, but the history of corporatism - and more specifically environmental precedents in the USA point to problems in implementation. However, the process might also be depicted as a *direct competition* between the groups for media and public opinion sympathy. While the oil industry is keen to talk to the environmentalists this may signify an attempt to persuade the environmentalists, rather than a new acceptance of environmental considerations within the decision processes of industry. The terms of debate have entered the fuzzy end of social science and moved away from the clarity (artificial perhaps) of 'hard' cost benefit decisions. Dealing with the 'wild animal' of the environmental consequences of decommissioning has raised questions about social decision making and processes that may be less easily solved than the technical difficulties discussed in other chapters. The social sciences may have some success in describing, but little in shaping, events.

References

1. Jordan, G. (1991) *The Commercial Lobbyists*, Aberdeen: AUP
2. Stewart, J. (1958) *British Pressure Groups*, Oxford: OUP
3. Truman, D. (1951) *The Governmental Process*, NY:Knopf
4. Eckstein, H. (1960) *Pressure Group Politics: the Case of the BMA.*, Stanford: Stanford U P.
5. Lindblom, C. (1977) *Politics and Markets*, NY: Basic Books
6. Connelly, P.C. (1992) *Dealing with Whitehall: A Practical Guide to Understanding and Influencing Government Decisions*, London: Century Business
7. Heclo, H. (1978) 'Issue Networks and the Executive Establishment', in A. King (ed), *The New American Political System*, Washington: AEI.
8. Rucht, D. (1995) 'Ecological Protest as Law-Breaking: Greenpeace and Earth First! in Comparative Perspective', in Rüdig, W. *Green Politics Three*, Edinburgh: Edinburgh University Press
9. Pearce, F. (1991) *Green Warriors: The People and the Politics Behind the Environmental Revolution*, London: The Bodley Head
10. Jordan, G. and Maloney, W. (1997) *The Protest Business?*, Manchester: MUP
11. Peters, G. and Barker, A. (1993) *Advising West European Governments*, Edinburgh: Edinburgh UP
12. Baumgartner, F. and Jones, B (1993) *Agendas and Instability in American Politics*, Chicago: Chicago U. P.
13. Graber, D. (1997) *Mass Media and American Politics*, 5th edn., Washington: Congressional Quarterly

Notes

i. In an interview in the *Sunday Times* in May (26th) 1996 he had admitted that Shell was culturally insular.

ii. UK00A estimate that the total cost of decommissioning in the UK sector will be £5bn with £1.6bn on the first 50 installations due to be removed by 2006.

iii. Of course there is a difference between 'being active' (even if one is influential) and being seen to be active in a way that will impress supporters.

iv. The London Convention was signed as an anti marine pollution device in 1972 as a development of the UN Conference on the Human Environment in Stockholm. Until 1992 this was called the London Dumping Convention but the name was changed to try to get rid of the impression that this was a pro dumping club.

v. The Oslo Convention will be superceded by the OSPAR Convention (annex 111).

vi. Greenpeace later (9th June 1995) revealed a report to Shell by Smit Engineering that claimed that the Brent Spar could be safely dismantled on shore at a cost of no more than £10 million. Greenpeace claimed that Shell suppressed this information and kept it out of its 'Best Practical Environmental Options and Abandonment Hypothesis'. On 20th June they were able to publish a memo from within the Fisheries Laboratory in MAFF dated 6th December 1993 what said the contents of the Brent Spar could not be dumped because of the toxicity. The chief scientist had annoted the memo to say that the waste 'cannot be dumped at sea' the only option is to 'take ashore and treat'. In other words the case for dumping was not one-sided among those with a specialist background.

vii. Press Release 9th June. Of course the Brent Spar was intended for the Atlantic rather than the North Sea.

viii. First report was 18th October. A new study reported on 30th November qualified the earlier report (further in the Shell direction) by noting that the Brent Spar did not contain PCBs.

ix. In an article in *The New Scientist* (23/30 December 1995), Peter Melchett made the important point that in Greenpeace's view their campaign was not an attack on science but about the limits of science - the uncertainties at the boundaries of science.

x. This was set up after Tim Eggar requested the Chief Executive of the Natural Environment Research Council to look at the disposal of large offshore structures. It was chaired by Professor John Shepherd, Director of the Southampton Oceanography Centre.

xi. The precise number of contractors and proposals varies slightly in different accounts.

Subject Index

MIX
Papier aus verantwortungsvollen Quellen
Paper from responsible sources
FSC® C105338

If you have any concerns about our products,
you can contact us on
ProductSafety@springernature.com

In case Publisher is established outside the EU,
the EU authorized representative is:
Springer Nature Customer Service Center GmbH
Europaplatz 3, 69115 Heidelberg, Germany

Printed by Libri Plureos GmbH
in Hamburg, Germany